River Logic

How Water Networks Carve Order From Chaos

Team Trex

Published by NobleTrex Press

For permissions and other inquiries, write to:

P.O. Box 3132, Framingham, MA 01701, USA

Contents

Introduction

If you stare out the window of an airplane at thirty thousand feet, the chaotic clutter of humanity fades away. The gridlocked highways, the angular subdivisions, and the sharp boundaries of property lines dissolve into a grey-brown texture. But if you look closely at the landscape below, a different kind of order reveals itself-one that is ancient, persistent, and stunningly intricate.

You see it in the silver threads winding through the dark green of a forest, or the stark, darker veins etched into the tan skin of a desert. You see it where water has carved a path for itself. From this height, the Mississippi River doesn't look like a shipping lane or a political border. It looks like a lightning bolt frozen in mud. It looks like the veins in a leaf, or the branching arteries of the human lung.

This resemblance is not an accident. It is a signature. It is the visible handwriting of physics solving a complex problem: how to move matter and energy from one place to another with the least amount of resistance.

We tend to think of rivers as static lines on a map-blue ribbons that politely stay within their banks, existing solely to fill our reservoirs or define our state lines. We treat them as plumbing. But a river is not a pipe. A river is a living, thrashing, decision-making entity. It is a dissipative structure, a thermodynamic machine that converts the potential energy of rainfall and gravity into the kinetic work of shaping the crust of the Earth.

This book is an exploration of that machine. It is an investigation into the hidden rules that govern how a raindrop falling on a mountain peak finds its way to the ocean, and how, in doing so, it collaborates with trillions of other drops to build a landscape that is mathematically predictable, geometrically beautiful, and terrifyingly powerful. This is the story of River Logic.

***

To understand this logic, we have to let go of the idea that water is formless. We are taught that liquids take the shape of their container, but in the wild, water *builds* its own container.

Consider the sheer improbability of a drainage basin. When rain falls on a jagged, chaotic hillside, it doesn't flow in a uniform sheet. It organizes. Droplets coalesce into rivulets, rivulets merge into brooks, brooks crash into streams, and streams swell into rivers. This process of aggregation is relentless. As recognized by pioneers of geomorphology like Robert Horton and Arthur Strahler, this organization follows a strict hierarchy. There is a mathematical architecture to the way tributaries join, a branching code so consistent that you can predict the number of streams in a basin just by measuring the main trunk.

This self-organization is the river's way of optimizing its existence. Nature is lazy; it hates wasted effort. A river network arranges itself in a way that minimizes the energy required to drain the land. It creates a fractal geometry-a pattern that repeats itself at every scale. You can see the same branching structure in a muddy roadside ditch that you see in the Amazon basin. The physics are identical; only the magnitude changes.

But water is only half the story. If rivers were merely water sliding over hard rock, they would be simple chutes. The complexity-and the chaos-arises because the river is armed. As it flows, it picks up tools. It carries silt, sand, gravel, and boulders. This sediment turns the flow into liquid sandpaper. The river uses this grit to slice through mountain ranges, carving canyons and widening valleys.

There is a delicate, unspoken negotiation between the water and the dirt. If the water moves too fast, it scours the bed, picking up more sediment than it can carry. If it moves too slow, it chokes, burying itself in its own debris. The river is constantly adjusting its slope, its width, and its depth to maintain a balance between the energy it possesses and the burden it must carry. This search for equilibrium is what makes rivers wind. It is why they meander in those mesmerizing, serpentine loops. A straight line might be the shortest distance between two points, but for a river trying to balance its energy, a curve is often the most efficient path.

***

Why does this matter to us, the species that has built its civilization on the banks of these networks?

It matters because our "Human Logic" is often at direct odds with River Logic. For five thousand years, we have settled on floodplains-the flat, fertile lands created by the river's need to breathe. We are drawn to them for their rich soil and easy transport. But we demand stability from a system designed for dynamism. We want the river to stay in one place, so we armor its banks with concrete and hem it in with levees. We build dams to stop the pulse of the seasons, turning dynamic flows into stagnant pools.

We treat the river network as a conquered beast, but the logic of the network is patient. When we straighten a channel to make shipping easier, we inadvertently steepen the slope, increasing the water's speed and its hunger for sediment. The river retaliates by eating away at the foundations of our bridges. When we wall off the floodplain to build subdivisions, we deny the river its natural shock absorbers. The energy that used to dissipate harmlessly across a marsh is now funneled downstream, turning a moderate high-water event into a catastrophic deluge for the next town on the line.

We are currently witnessing a global collision between the static world we have built and the dynamic world of water. In the Mississippi Delta, the land is sinking and vanishing into the Gulf of Mexico because our engineering has starved the coast of the sediment it needs to survive. In Europe and Asia, "100-year floods" are arriving every decade, fueled by a changing climate that is injecting more energy into the hydrologic engine than our infrastructure was designed to handle.

To navigate the coming century, we need a fundamental shift in perspective. We cannot simply engineer our way out of these problems with more concrete. We have to learn to think like a river.

***

Our inquiry starts with the source of the motion itself: the engine of gravity. Before there is a river, there is potential energy. The hydrologic cycle is essentially a giant planetary pump, powered by the sun, lifting ocean water into the sky and dropping it on the continents. Once that water hits the ground, it has only one goal: to return to base level, usually the sea. The path it takes is a negotiation with geology.

We will trace this journey from the headwaters, where the energy is high and the landscape is raw. Here, we see the river as a knife, cutting vertical notches into the bedrock. We will watch how the network assembles itself, obeying the scaling laws discovered by twentieth-century hydrologists who realized that nature's chaos has a predictable rhythm. We will see how these networks are not random tangles, but highly efficient systems that have evolved to drain the continents as quickly as friction will allow.

As the river leaves the mountains, the story shifts from erosion to transport. We will explore the mechanics of the "sediment conveyor belt." This is where the river becomes a builder. We will look at the physics of the meander, examining how a slight perturbation in flow creates a feedback loop that causes a river to swing wildly across a valley floor, leaving behind oxbow lakes and scars of its past positions. This wandering is not aimless; it is the mechanism by which the river constructs its own floodplain.

Understanding the floodplain is crucial because it challenges our definition of what a river is. We tend to think the river ends at the water's edge. River Logic suggests otherwise. The floodplain is not *land next to the river*; it is the *river in storage*. It is a temporary parking lot for water and sediment. When we forget this-when we pave over the sponge-we break the hydraulic circuit.

Finally, the network reaches its conclusion at the sea. Here, the logic reverses. The flow slows, the energy dissipates, and the river drops its burden. It builds a delta-a fan of new land that pushes back against the ocean. This is the most fragile part of the system, a place where the balance between the river's construction and the ocean's destruction plays out in real-time. It is also where the conflict with human engineering is most acute.

***

Writing about rivers is, in some ways, an exercise in humility. The systems we are discussing are so vast and complex that they defy simple narratives. A single river is a hydraulic machine, a biological corridor, a geological agent, and a cultural artery all at once. It obeys the laws of fluid dynamics, but it also behaves with a strange kind of agency, correcting itself, healing its wounds, and seeking balance.

Throughout this exploration, we will meet the scientists who cracked the code of these networks. We will look at the work of Luna Leopold, the son of the famous conservationist Aldo Leopold, who brought rigorous physics to the study of scenic streams and showed that a river's shape is a quantitative certainty, not a poetic accident. We will look at the concept of "grade"-the theoretical state of perfection every river strives for-and how a river will cannibalize its own bed to reach it.

But this is not just a book about history or physics; it is a user's manual for a planet in flux. As we face a future of intensified storms and rising seas, the old strategy of command and control is failing. We are seeing levees breach and dams crumble. But a new philosophy is emerging, one that focuses on "room for the river." By understanding the inherent logic of the water network, engineers and ecologists are beginning to work *with* the flow rather than against it-removing dams, reconnecting floodplains, and allowing the river to reclaim some of its ancient, winding freedom.

The shapes on the map are not fixed. They are frames in a movie that is running too slowly for us to see. But if we learn to read the water, we can see the next few frames before they happen. We can understand why the river creates a sandbar here, why it erodes a bank there, and why it will inevitably retake the land we thought we owned.

So, look out the window. Look at the ditch by the highway, the creek behind your house, or the great grey expanse of the river running through your city. They are all doing the same thing. They are all part of the same vast, branching equation, solving for gravity, carving order from chaos. Let's see how they do it.

Chapter 1

The Gravity Engine

A drop of water at the Continental Divide possesses a quiet, pent-up energy waiting to be unleashed. This chapter uncovers how solar heat and tectonic lift conspire to build a planetary engine, turning the static weight of the atmosphere into the relentless, carving force of a river in motion.

1.1 The Atmospheric Conveyor

Rain has a way of sounding local. It drums on one roof, misses the next street, and turns a familiar sidewalk into a thin reflective sheet. Yet every drop that hits a continent is the end of a long, planetary journey. Before there is a stream in a gully, before there is a river in a valley, there is an act of lifting: water mass raised against gravity, stored as potential energy, and then handed back to the surface as a kind of down payment on future flow.

The common diagram of the hydrologic cycle can make it feel like a tidy loop: evaporation, clouds, precipitation, runoff, ocean. Neat arrows, neat labels. The real story is less tidy and more muscular. The arrows represent not only movement of water but movement of energy. Sunlight, absorbed mostly at the ocean surface, pays the energetic cost of turning liquid water into vapor. That cost is large and specific: the latent heat of vaporization, the hidden energy that breaks water molecules free from the liquid's cohesion. Once in the air, that energy is not lost; it is carried invisibly with the vapor, like a receipt tucked into a pocket. When vapor condenses into cloud droplets and raindrops, the receipt is cashed: latent heat is released to the atmosphere. This is one reason storms can organize and intensify. Condensation does not merely make clouds; it warms the air column and helps keep it buoyant, shaping winds that can draw in more moisture.

Seen this way, a river begins with a global distillation system. The oceans are the boiler plates. The atmosphere is the moving belt that carries the distilled product inland. Mountain ranges are the cold coils where vapor is forced to condense. Continents are the slanted trays where gravity takes over, pulling water back toward the sea. The cycle is not a circle drawn on paper but a set of coupled engines, imperfect but relentless: one powered by solar radiation and phase changes, another powered by gravity and the slope of land.

The "lifting" is literal. Water that evaporates from sea level and later falls as snow on a high plateau has been raised by kilometers. A kilogram of water lifted by 3,000 meters gains gravitational potential energy of roughly

$$E \approx mgh \approx (1\ \mathrm{kg})(9.8\ \mathrm{m/s^2})(3{,}000\ \mathrm{m}) \approx 29{,}000\ \mathrm{J}.$$

That energy is modest compared with the latent heat that was required to evaporate it (millions of joules per kilogram), but it is the form of energy that rivers can spend. It is the part of the atmospheric transaction that ends up as the kinetic force of flow, the ability to push sediment, to wear rock, to flood a plain, to carve a channel and reshape a coastline. The atmosphere is not only delivering water; it is loading the landscape with elevated mass.

The conveyor has a geography. There are places on Earth where the air is, on average, taking water away (net evaporation) and places where it is giving water back (net precipitation). That imbalance is not a bookkeeping error; it is the reason rivers exist at all. A continent cannot keep receiving rain unless something somewhere is losing water to evaporation. The simplest way to grasp the scale is to remember that river discharge into the oceans is not a trickle but a planetary flux. To supply it, the atmosphere must continually export water vapor from evaporating regions to precipitating regions. That export is accomplished by winds organized on scales from daily sea breezes to global circulation cells.

The engine driving those winds is uneven heating. The tropics receive more solar energy than the poles; the planet's rotation then helps sort the resulting motions into bands and storms. Add water vapor and the whole machine changes character. Dry air rising and cooling is one kind of buoyancy story; moist air rising and condensing is another, because condensation releases heat exactly where it matters most: in the rising column. The atmosphere can be thought of as a heat engine that converts some of the imposed temperature contrasts into mechanical motion, but with moisture acting as both fuel and complicating impurity. A humid atmosphere transports energy poleward in the form of latent heat and also modifies how efficiently mechanical work can be extracted from those temperature differences. The details belong to atmospheric physics, but the consequence is familiar: when the sky is moving water, it is also moving power

This is why rivers have a meteorological ancestry. The character of a river basin—whether it tends to receive water in steady doses or in violent pulses—is often set long before the water reaches a hillside. A winter storm track that repeatedly grazes a coast will build one kind of hydroclimate; a monsoon that arrives like a door opening and closing will build another. The atmosphere is the upstream.

Consider a concrete case: California's dramatic winter floods and, in between, its long dry spells. Much of the state's annual precipitation can arrive in a handful of events known as *atmospheric rivers*—long, narrow corridors of concentrated water vapor transport, often embedded in the wind fields of midlatitude cyclones. The name is metaphorical but the physics is literal: these filaments can carry more water per unit time than the flow of many of the world's great terrestrial rivers. When one of these corridors makes landfall and is forced to rise over the Coast Ranges and the Sierra Nevada, vapor condenses rapidly and precipitation intensifies. The same mechanism that stocks mountain snowpack can, in a warm storm, release rain onto already wet soils and swollen channels. In a few days, landscapes can experience the kind of runoff delivery that would otherwise be spread across months.

The atmospheric river story also contains an older human narrative about measurement and surprise. In the nineteenth century, as Western states and territories argued over water rights and tried to make agriculture reliable in dry climates, they learned that averages could be misleading. A mean annual rainfall number, printed in a survey report, did not convey the fact that much of the year could be rainless, and then a sequence of storms could arrive so intensely that levees failed and fields turned into lakes. California's Great Flood of 1861—1862, for example, was produced by a series of winter storms that inundated the Central Valley for weeks, transformed Sacramento into a watery grid, and effectively relocated the state government for a time. The physical trigger was meteorological, but the lesson was hydrologic and political: a region's water supply and its flood hazard can be delivered by the same sky mechanism, depending on timing and temperature.

That duality—benefit and hazard bundled together—is a hallmark of the conveyor. Water delivered slowly is storage. Water delivered fast is force. A storm that falls as snow can be a delayed release, stored in mountain reservoirs made of ice crystals. A storm that falls as rain on snow can be a sudden liquidation, flushing meltwater into channels at the same time rainwater arrives, producing floods that look disproportionate to the rainfall alone. The atmosphere decides the phase and the pace; the basin responds.

Mountains, in turn, are not passive recipients. They are active components of the conveyor because they convert horizontal vapor transport into vertical motion. When moist air encounters high topography it must rise (or flow around, if it can). Rising air expands and cools; cooling pushes it toward saturation; saturation forces condensation and precipitation. This is *orographic* precipitation—rain and snow "wrung out" by relief. On the windward side, forests can live and rivers can be fed; on the leeward side, the same range can cast a rain shadow where deserts form. The basic geometry is familiar to anyone who has driven across a mountain pass and watched the vegetation change as if someone had switched biomes at a boundary line.

The orographic effect also helps explain why so many great rivers begin in high places that seem, at first glance, too cold or too barren to be major water sources. The headwaters of large basins are often not where most of the basin's area lies, but where the precipitation is most reliably converted into runoff: high elevations where snow persists, where summer temperatures are mild, where evapotranspiration (the combined loss of water by evaporation and plant transpiration) is limited by energy rather than by available moisture. A mountain range is, in this sense, a condenser and a savings account. It condenses vapor into snow and stores it in a form that resists immediate return to the atmosphere.

The savings account is never perfectly secure. Snow sublimates; glaciers melt; winds redistribute drifts; warm spells do their quiet work. Still, the seasonal rhythm of snow accumulation and melt is among the most consequential timing devices on the planet. It transforms a winter-heavy precipitation regime into spring and summer river flow, aligning water supply with growing seasons and human demand in many temperate regions. Where that alignment is relied upon—in the Indus, the Colorado, the Po, the Rhône—a warming climate does not merely change "how much" water arrives. It changes when water is released from storage, and whether it is stored as snow at all.

One can treat the hydrologic cycle as a mass balance: precipitation in, evapotranspiration out, runoff as the remainder. That accounting is necessary, but it hides a deeper point: the atmosphere delivers water to land as *organized inequality*. Some places get more than they lose; other places lose more than they get. Rivers are the drainage system that resolves that inequality by returning the surplus to the sea. In other words, a river's existence is proof that, for its basin, precipitation has exceeded evapotranspiration often enough to build a persistent outflow. Where that condition fails, drainage networks become ghostly: channels may exist and even flash with water after storms, but there is no perennial river, only episodic flow and evaporative disappearance.

The conveyor is also selective about where it puts its gifts. A storm does not distribute water evenly across a catchment. It paints the landscape with gradients and patches. The spatial pattern matters because the first question for a river is not "How much water fell?" but "How much of what fell becomes connected flow?" Two hillsides receiving the same rainfall can produce very different runoff depending on soils, vegetation, and antecedent wetness, but the atmosphere has already loaded the dice by deciding where the rainfall is intense, where it is gentle, where it falls as snow, and where it evaporates before reaching the ground.

There is a small and easily overlooked step in the sky-to-river story: much precipitation never reaches a channel at all. It is intercepted by leaves and evaporated. It infiltrates into soil and is later transpired by plants. It fills depressions and disappears in the next dry spell. Even in wet climates, only a fraction of precipitation becomes streamflow on the timescale we notice. The rest cycles locally between land and air. That local cycling is not a rounding error. In many regions, it is the main form of exchange, and it makes the atmosphere and the biosphere partners in deciding how much water is available to become runoff.

The biosphere's involvement can be visceral. Walk into a forest after rain and the air feels thick, not only with humidity but with a scent that seems to announce that the trees and soil are exhaling. Plants are not passive straws. They open and close stomata, trading water for carbon dioxide, regulating loss in response to light, heat, and stress. This means that the river network downstream is, in part, reading the physiological state of vegetation upstream. During drought, even when a storm arrives, a parched landscape can absorb and retain water differently than a wet one; the first rains can vanish into dry soils and thirsty roots, producing little immediate flow. The conveyor delivers, but the catchment decides how much is spent as discharge and how much is reinvested as storage.

Still, the most dramatic conversion from sky water to river water happens when the atmosphere delivers moisture faster than the land can hold it. The geometry of storms matters here. A slow moving storm that repeatedly trains over the same region can produce exceptional totals. A convective outbreak can dump intense rainfall over a small area, producing flash floods in tiny basins while leaving the larger river only mildly perturbed. In mountainous terrain, a modest shift in wind direction can focus precipitation on one tributary rather than another, changing which valley floods and which remains quiet.

This is why flood histories can feel so capricious to those who live in valleys: the triggers are often mesoscale patterns in the atmosphere that do not respect political boundaries or human expectations.

The conveyor's most important gift to rivers, though, is not the drama of extremes but the steady replenishment of elevated water mass. Every year, on average, water is lifted from the ocean and redeposited on continents at higher gravitational potential. If one wanted a single image to hold in mind, it would be a global pump that charges countless small batteries. Each mountain snowfield and upland aquifer is a charged battery. Each saturated soil layer in a wet season is a charged battery. Rivers are the circuits that discharge those batteries, sometimes slowly, sometimes in sparks.

There is a temptation to picture gravity as the only actor that matters once water is on land. Gravity does become the direct driver of downslope motion, but the atmosphere has already done a subtler job: it has decided where on the surface the water will begin its descent. A raindrop that falls on a ridge a meter from a divide might end up in one ocean basin; a raindrop that falls a meter the other way might end up in another. The atmosphere's placement of precipitation relative to divides is a form of routing that happens before any channel is involved. It is also why changes in storm tracks can have outsized impacts on water resources even if the total amount of precipitation over a large region changes only modestly. Shift the delivery upstream or downstream, windward or leeward, and the effective supply to particular reservoirs and rivers can swing sharply.

Historical hydrology offers an instructive anecdote from the age when engineers and surveyors first tried to make the sky legible to arithmetic. In the mid-nineteenth century, Britain's George James Symons helped organize systematic rainfall measurement, building networks of observers and standardizing gauges. The motivation was not purely scientific. Cities needed dependable water; canals and railways needed planning; floods demanded explanation. Early observers learned quickly that rainfall could vary dramatically over short distances, especially in hilly terrain. A single gauge could not stand for a whole district. This was not merely inconvenient; it was a clue. The atmosphere was not sprinkling water uniformly but delivering it in structured patterns shaped by wind, topography, and storm dynamics. Modern radar and satellites have made that structure visible in real time, but the lesson was there in the notebooks: the sky's conveyor belt has seams and knots.

It is also worth noticing how the conveyor ties rivers to distant oceans. Many of the storms that feed western North America draw moisture from the subtropical Pacific; storms feeding Western Europe can tap the Atlantic; monsoons are organized by land—sea temperature contrasts across entire basins. When a river runs high, it can be responding to conditions far offshore weeks earlier: anomalously warm sea surface temperatures that enhance evaporation, a shifted jet stream that steers storms, a blocking pattern that stalls them. To treat a river as merely a feature of its valley is to miss its long reach into the atmosphere.

At the end of the chain, the remarkable part is how ordinary the result can look. A creek that sings over stones in late spring feels intimate, as if it belongs only to that slope and that patch of shade. Yet the creek is the visible discharge of a hidden and global transaction: sunlight used to pry water from the ocean, winds and condensation used to carry it inland and aloft, mountains used to squeeze it into rain and snow, and gravity used to spend its elevated mass. The water's descent is the last step, but the river's beginning is written in the sky, in the invisible lanes of vapor transport that continually reload the continents with the raw material of flow.

1.2 The Geometry of Descent

Once water has been lifted and released onto land, gravity takes over with the single-mindedness of a law. It does not negotiate. A raindrop on a ridge has one reliable future: down. Yet the path from crest to coast is rarely a straight fall. Rivers do not simply accelerate until they become torrents; they develop a recognizable shape, a long, smooth curve in profile, steep at the top and gentle at the bottom. That curve is the landscape's way of putting brakes on gravity's constant pull.

A useful way to picture a river is not from above, as a blue line on a map, but from the side, as if the valley were cut open and you could look along the channel from its headwaters to its mouth. Plot elevation on the vertical axis and distance downstream on the horizontal axis. The resulting line is the river's *longitudinal profile*. For many rivers, especially those that have had time to adjust to their setting, that line is *concave-up*: steep near the source, progressively flattening toward the sea. The shape is so common that it can feel inevitable, like a signature of nature. It is not inevitable. It is the outcome of a negotiation between forces that pull water downhill and forces that resist it.

The pull is gravity, but gravity is not a force that acts on the river alone; it acts on everything the river carries and everything the river tries to move. A river's descent is a conversion of gravitational potential energy into other forms: the kinetic energy of flowing water, the turbulence that mixes air and water, the heat dissipated by friction, the work of breaking rock and transporting sediment. If the conversion were perfectly efficient, a river would be a kind of conveyor belt for mechanical work. It is nothing of the sort. Most of the potential energy is lost as heat through friction at the bed and banks, and through internal turbulence. The work a river accomplishes on the landscape comes from a small but persistent remainder, applied day after day, flood after flood, over spans of time that are hard to hold in mind.

Friction is not merely an annoyance; it is the key to why rivers have shape. Water sliding along rock and gravel is resisted by roughness. Flowing around bends and over boulders sheds energy into eddies. Sediment in the water increases effective viscosity and changes turbulence. The channel itself is not a smooth chute; it has a bed made of particles and steps, it has pools and riffles, it has vegetation and woody debris. Every one of these features takes a little bite out of the energy of descent. A river profile is, in this sense, a record of where energy is being spent.

The simplest way to see the logic is to think in terms of *slope*, the drop in elevation per unit distance downstream. Slope matters because it sets how much potential energy is available to be converted into motion over that distance. Steeper slope gives the river more energy to spend; gentler slope gives less. But slope is not chosen freely. It adjusts, over time, in response to two stubborn constraints: what the river must accomplish, and what it is allowed to do.

What it must accomplish is the export of water and sediment out of its basin. Water arrives from the sky; sediment arrives from weathering, landslides, bank collapse, and the slow creep of hillslopes. Sediment is not optional. Even in bedrock gorges where solid rock seems to dominate, rivers tend to carry a mobile layer of stones and sand during high flows, and that mobile layer can act like sandpaper or like armor. A channel that is choked with sediment behaves differently from a channel that is mostly bare rock. The river must carry what it is given, and that requirement affects the slope it needs.

What it is allowed to do is set by its *base level*, the lowest elevation to which it can erode. For most rivers, the ultimate base level is sea level, because the ocean is the terminal reservoir. But base level can be local as well: a lake, a resistant bedrock ledge, a downstream dam of ice or debris, even the bed of a larger river into which a tributary flows. Base level is a kind of floor. A river can cut down toward it, but not below it, and changes in that floor propagate upstream as changes in slope.

These ideas can sound abstract until you notice that they are present in ordinary experience. Imagine standing at the mouth of a river where it meets the sea. The channel is broad; the water surface is nearly flat; the current may even appear sluggish. Walk upstream into the foothills and the water begins to talk louder. The channel narrows, the bed becomes coarser, the gradient steepens, and the flow becomes visibly energetic. Farther up, in the headwaters, flow may tumble over steps, plunge through boulder cascades, and in places disappear into porous debris. The same river, the same gravitational field, but different slopes, different sounds, different abilities to move material. The profile ties those differences together into one continuous curve.

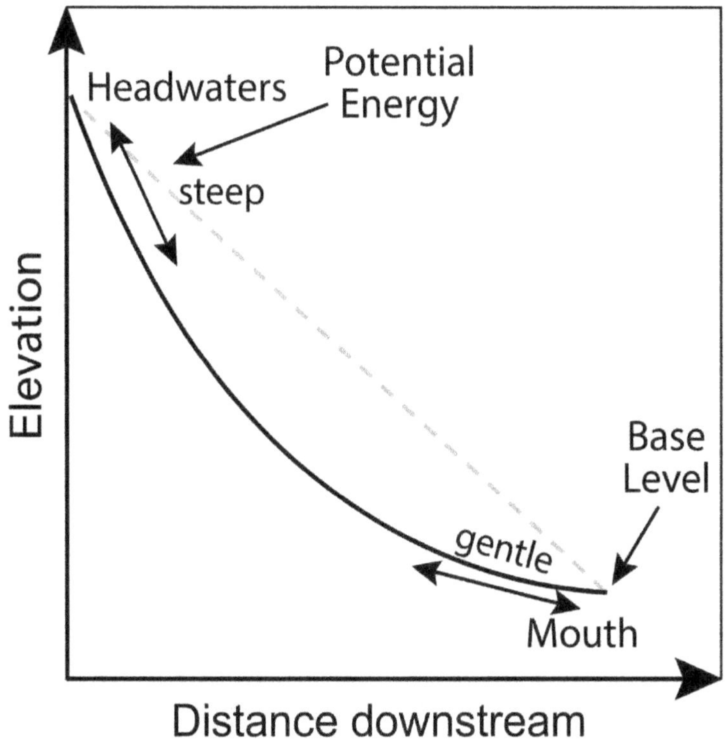

The concave shape is not merely aesthetic. It reflects a scaling property of drainage basins: as you move downstream, more tributaries join, the *drainage area* contributing water increases, and typical discharges become larger. A larger river can do the same geomorphic work at a lower slope because it has more water power available. This is the intuitive heart of the concave profile: small headwater streams often need steep gradients to mobilize coarse material and to cut into rock, whereas large lowland rivers, with far greater discharge, can transport huge sediment loads across gentle slopes.

People sometimes imagine the headwaters as delicate beginnings and the lowlands as the realm of power, because the lowland channel looks large. In terms of the landscape's resistance, it is often the reverse. Headwater channels may be the places where the river must work hardest per meter of downstream distance, because they face coarse debris and resistant bedrock with relatively little water. That is why mountain streams can look overbuilt: steep, loud, and armored with boulders. They are the places where the river spends a large share of its potential energy budget quickly.

A natural question follows: if downstream rivers have more water, why doesn't every river simply flatten out everywhere? The answer is that a river's slope cannot be arbitrarily reduced, because it still must maintain the ability to transport the sediment delivered from upstream and hillslopes. If slope becomes too low for the available flow conditions, sediment accumulates, raising the bed, changing the channel geometry, sometimes forcing the river to braid into multiple threads. If slope becomes too steep relative to what the channel can resist, the river will erode, incise, and lower its bed. This self-adjustment is one reason river profiles often appear smooth even though they are built from violent events. Over long periods, the channel tends to find a profile where its capacity to erode and transport roughly balances the supply of sediment and the imposed base level.

It helps to introduce a simple measure of the rate at which a river can do work on its bed. Geomorphologists often talk about *stream power*: the rate of potential energy loss per unit length of channel. In its most basic form, total stream power scales like

$$\Omega \approx \rho g Q S,$$

where ρ is water density, g is gravitational acceleration, Q is discharge (volume flow rate), and S is slope. Even without diving into details, the expression tells a story: bigger flows and steeper slopes both increase the river's capacity to expend energy on erosion and transport. Because Q typically increases downstream, S can decrease while maintaining comparable stream power. A concave profile is a geometry that spreads energy expenditure across the river's length in a way that keeps the system dynamically plausible.

Yet there is an additional complication: rivers do not flow at an average discharge; they flow as a sequence of low flows punctuated by floods. Much of the sediment transport and much of the erosive work occurs during relatively short, high-discharge events. A profile that appears stable is, in fact, an emergent average over countless storms. The brakes are applied intermittently, and the accelerator is pressed hardest during floods.

Base level provides the other major lever. When base level drops, a river suddenly has more vertical distance to fall. It responds by steepening in some reach, increasing its capacity to erode downward, and the resulting change migrates upstream. When base level rises, the opposite occurs: the river's lower reaches lose the ability to cut down, sediment tends to accumulate, and the river may aggrade, building its bed upward and flattening its profile near the mouth. Sea level changes during ice ages are a classic global driver. When ice sheets lock up water on land, sea level falls, and rivers can incise into their own valley fills, leaving terraces. When sea level rises again, valleys may fill with sediment and become estuaries.

Not all base-level changes come from oceans. Consider a river encountering a hard rock ledge. Upstream of the ledge, the channel may flatten and deposit sediment; downstream, it may plunge and scour. Such a ledge creates a *knickpoint*: a localized steep step in the longitudinal profile. Knickpoints can also form when a fault uplifts a block of land, when a landslide dams a valley, when a glacier leaves a moraine as a temporary barrier, or when a river is captured by a neighboring basin and its outlet elevation changes abruptly. The profile, in these cases, is no longer a smooth concave curve but a curve interrupted by a kink. The kink matters because it concentrates erosive energy and it carries information: it is a scar of change, a sign that the river is still adjusting.

A concrete example can be found in the landscape around Niagara Falls. The falls are often treated as a tourist spectacle, but they are also a geomorphic signal of base-level adjustment. After the last glaciation, meltwater and reorganized drainage routes established an outlet from the Great Lakes toward the Atlantic. Niagara's lip is a knickpoint maintained partly by the contrast between resistant caprock and softer rocks beneath. The falls have migrated upstream over time as erosion undercuts the lip. The dramatic drop is not simply a feature pasted onto an otherwise gentle river; it is a moving disturbance in the longitudinal profile, a place where the river spends potential energy rapidly to erase a mismatch between upstream elevation and downstream base level.

If Niagara is a postcard example, many knickpoints are subtle: a reach that is steeper than expected, a short gorge, a set of rapids where the channel narrows and the bed becomes rough. To a paddler, these are hazards; to a geomorphologist, they are punctuation marks. They imply that the river is not in a steady balance everywhere at once. Rivers remember changes in uplift, climate, and base level as transient shapes in their profiles.

The long profile also ties rivers to tectonics, the slow deformation of Earth's crust. In mountain belts, uplift raises the land, creating new potential energy. Rivers respond by incising, cutting down through rock. This incision, sustained over millions of years, can help limit how high mountain ranges become: as uplift builds relief, rivers carve it down. The landscape becomes a tug-of-war between rock pushed up and rock carried away. Bedrock rivers are central players in that contest because they often provide the only efficient routes for lowering the interior of rising ranges. Their channels can be narrow and steep, their flow concentrated, their erosive work intense during rare floods. The shape of a river profile in such settings is not just a consequence of gravity; it is a record of how rapidly the land is being built and how effectively water can dismantle it.

The historian's version of this story is written in the development of ideas about landscape evolution. In the late nineteenth century, the American geologist William Morris Davis proposed a descriptive picture of how landscapes might pass through stages from youthful ruggedness to mature gentleness. Many details of that scheme have since been revised, criticized, and replaced by more process-based thinking, but the motivation remains familiar: landforms seem to organize themselves into patterns, and rivers are among the most legible of those patterns. Davis and his successors were, in effect, trying to read the long profile as a sentence about time. Modern geomorphology is more cautious about neat stages, but the instinct to read rivers as narratives persists because it is often correct in limited, testable ways: a knickpoint can indicate a change; a concavity can reflect how discharge grows downstream; terraces can indicate episodes of incision and filling.

For a more intimate example, imagine a small river in a mountainous region with a dam built in the twentieth century. The dam creates a new local base level upstream: a reservoir surface that is flat and fixed by engineering. Sediment that would have traveled downstream is trapped. Upstream, the channel may aggrade as sediment accumulates in deltas at the reservoir's head. Downstream, the released water can be sediment-starved and therefore unusually erosive, hungry for grains to carry. It may incise into its own bed, undermining banks, coarsening the channel bed as finer material is swept away. The longitudinal profile is re-edited by a single structure. If the dam is later removed, another disturbance propagates: stored sediment can flush downstream, channels may temporarily braid, and the river begins the slow work of re-establishing a profile compatible with its new base level and sediment regime. The shape is not an ornament; it is the river's response to constraints.

The phrase "the landscape sets the brakes" is not just metaphor. There are physical mechanisms that act as braking systems. One is simple friction, intensified by roughness. Another is sediment itself, because moving sediment consumes energy. A third is channel width and depth, which adjust so that shear stresses on the bed are sufficient to move material during high flows but not so extreme that banks fail continuously. A fourth is bedrock strength: some lithologies resist incision more than others, requiring steeper slopes or higher discharges to cut. Even vegetation plays a role by stabilizing banks, adding roughness, and influencing floodplain formation. The river profile is the integrated outcome of all these brakes interacting with the gravitational drive.

At a dinner table, someone might ask why the river does not simply take the straightest line downhill. The answer is that "downhill" is not a single direction; it is a field of possible gradients across a rugged surface. Water initially follows the steepest local descent, but channels evolve. They incise where flow concentrates; they abandon paths when sediment chokes them; they migrate laterally across floodplains; they are deflected by resistant ridges and guided by structural weaknesses. Over long periods, the river network organizes so that most water, most of the time, follows routes that efficiently connect high potential energy sources to low base levels under the constraints of rock, sediment, and climate. The longitudinal profile is the backbone of that organization.

One last way to make the geometry emotionally tangible is to recall that people, too, feel slopes in their bodies. A steep mountain creek looks like danger because it is: slip in, and gravity will not wait. A lowland river looks calm, but that calm can be deceptive because the river's power is hidden in volume, not gradient. During floods, a broad low-slope river can carry enormous momentum, uprooting trees, lifting cars, and peeling asphalt from roads. The profile tells you where energy is stored and where it will be released, but it does not tell you when. That timing belongs to storms and seasons. Still, the shape of the descent sets the stage: it determines how quickly water can translate rainfall into motion, how much energy is available per meter of channel, and where that energy is likely to be spent.

Stand at the mouth of a river at low tide and pick up a rounded pebble. It may have started as a sharp fragment in a headwater landslide, high above the tree line, where the profile is steep and the water is cold. Its edges were knocked off in cascades, its surface polished by repeated collisions, its journey paused and resumed with floods. The pebble is a small, tactile record of the river's curve, of energy spent unevenly along the way: rapidly in the steep headwaters, patiently in the long lowland glide, and finally surrendered at base level where the river's ability to carry begins to fail and the sea takes over.

1.3 The Pulse of the Catchment

A river on a map looks like plumbing: a line that collects smaller lines and delivers everything to a single outlet. That view is useful for navigation and property boundaries, but it quietly suggests a mistake. Pipes are predictable. Give a pipe a steady input and it gives a steady output, with only a short delay. Rivers, even in the most orderly landscapes, behave more like living systems. They swell and shrink. They remember yesterday's rain. They react differently to the first storm after a drought than to the fifth storm in a wet month. They can go from ankle-deep to house-high in a few hours, then drop back as if the flood were a dream. The word for their moment-by-moment condition is *discharge*.

Discharge, usually written Q, is the volume of water passing a point per unit time. It is the accountant's unit of rivers: cubic meters per second, or, in older American practice, cubic feet per second. It is also the river's pulse. If you stand on a bridge and watch the current, you can feel it as speed; if you look down at the water level, you feel it as stage height. But Q is the deeper quantity, because it expresses how much of the catchment has become hydraulically connected to the channel network at that moment.

A *catchment* (or drainage basin) is the area of land that drains to a given point on a stream. Its boundary is a divide, often a ridge line. Everything inside that boundary is a potential contributor to flow at the outlet, but only some of it contributes at any one time. That "some" is not fixed; it expands and contracts with weather and season. The key to the catchment's pulse is that the basin is full of temporary storage: in leaves, in snowpacks, in soil pores, in fractured rock, in wetlands, and in groundwater. Storms do not simply pour into a channel; they are negotiated through storage.

The simplest mental model of rainfall and runoff is a bucket. Rain falls into the bucket; when it overflows, water becomes runoff. The model is not wrong, but it hides the basin's internal architecture. Real catchments are not one bucket but a ragged set of buckets connected by leaky tubes, with valves that open and close depending on how wet everything already is. Some stores fill quickly and drain quickly, like a thin soil on steep slopes. Others fill slowly and drain slowly, like deep groundwater in fractured bedrock. The same storm can produce a sharp spike in Q in one basin and a gentle swell in another because the internal stores and connections are different.

Hydrologists draw the pulse as a *hydrograph*: discharge on the vertical axis, time on the horizontal. A hydrograph has a rise, a peak, and a recession. The rise tells you how quickly the basin delivered water to the channel. The peak tells you the maximum connectivity achieved during the event. The recession tells you how the basin empties its stores back toward the channel after rainfall stops. Even people who have never used the word "hydrograph" are familiar with its implications. Flood warnings, reservoir operations, canoe trips, and insurance maps are all, in one way or another, bets on the shape of future hydrographs.

The most important thing a hydrograph teaches is that discharge is not simply "rain turned sideways." Between the sky and the channel lies the land's capacity to absorb, delay, and reroute water. That capacity is heavily state-dependent. When soils are dry, much rainfall infiltrates. When soils are saturated, rainfall is rejected and becomes surface runoff. When the water table is low, hillslopes can take in water like a sponge. When the water table is high, a basin can behave like a tabletop: water skates across it.

This state-dependence is why catchments can be so deceptive. A small storm can cause a big flood if it lands on a basin already primed with wet soils and high groundwater. A large storm can produce only modest streamflow if it falls on a basin that is thirsty and able to store. People sometimes frame floods as purely meteorological accidents, but many floods are as much about antecedent wetness as about rainfall intensity. The catchment does not merely receive the storm; it interprets it.

One of the most influential ideas in hydrology is the concept of *variable source areas*. Rather than assuming that runoff comes uniformly from everywhere in the basin, the variable source area picture says that only certain parts of the landscape generate quick runoff during a storm, and that the size of those parts changes through time. The classic mechanism is *saturation-excess runoff*, sometimes called the Dunne mechanism after Thomas Dunne's field-based work. The idea is straightforward. In many humid catchments, rainfall can infiltrate most hillslopes readily. Overland flow (water running across the ground surface) is not created because the rain falls too fast for the soil to absorb it, but because the soil has already reached saturation. When the near-surface soil is full, additional rainfall has nowhere to go but sideways, downhill, and quickly into a channel. Sometimes water also rises from below: the water table pushes up to the surface in hollows and valley bottoms, creating patches of saturation that can connect directly to streams. Those patches expand during wet periods and contract during dry periods.

This picture makes intuitive sense if you have walked in a valley after prolonged rain. Ridges remain firm underfoot, but low swales squish. Springs appear where yesterday there were none. A shallow depression becomes a mirror. These are the basin's lungs filling. When the rain continues, those wet patches merge into connected pathways. A catchment can suddenly "switch on," delivering a large fraction of rainfall to the stream with little delay, not because every square meter is producing runoff, but because a connected network of saturated areas has formed.

Topography is the organizer of that switching. Gravity pulls water downhill not only in channels but within soil and shallow groundwater. Concave hollows tend to accumulate subsurface flow; valley bottoms collect both water and sediment; steep convex ridges tend to shed water into deeper storage. This is why, in many landscapes, the wettest places are predictable even without seeing the rain. Hydrologists formalized this intuition in models that link topography to saturation tendency. TOPMODEL, developed by Keith Beven and Mike Kirkby in 1979, is a particularly elegant example. It proposes that the tendency for a point on the landscape to become saturated can be related to a simple index that combines two factors: how much land drains through that point (an upslope contributing area, often written a) and how steep the local slope is (often written as $\tan\beta$). The *topographic wetness index* (TWI) is commonly written as

$$\text{TWI} = \ln\left(\frac{a}{\tan\beta}\right).$$

Large a means a lot of upstream area sending water through; small $\tan\beta$ means gentle slope, which encourages accumulation and slows drainage. High TWI values often correspond to valley bottoms and convergent hollows where saturation is likely; low values correspond to steep ridges where water drains away.

Even without calculating it, you can see the logic. A steep hillside is like a tilted tray: water moves through and away. A broad, gentle hollow is like a shallow bowl: water lingers, rises, and spreads. The wetness index puts numbers to that qualitative picture. It has become a workhorse because it helps connect what you can measure easily from topography to something that is otherwise hard to map at scale: the spatial pattern of near-saturation, and therefore the likely pattern of fast runoff generation.

There is a twist, and it matters for how we interpret "maps of wetness." In practice, a is computed from digital elevation models using algorithms that decide how to route flow across a gridded surface. Different algorithms make different choices about whether flow is distributed across multiple downslope directions or forced into a single path, and about how to handle flats and depressions. Those choices can substantially change the resulting contributing area and therefore the TWI field. When people compare wetness index maps to field observations—soil moisture, groundwater level, vegetation indicators—the correlations can depend on method as much as on physics. The basin's pulse is real, but our attempt to draw it on a map carries its own uncertainty, and that uncertainty has consequences when such maps are used for predicting flood-prone areas, designing roads, or restoring wetlands.

The pulse is also shaped by where channels begin. A hillslope does not automatically have a channel; there is usually a threshold where diffuse flow becomes concentrated enough to carve a persistent channel head. David Montgomery and William Dietrich's work in the late 1980s helped frame channel initiation in terms of slope and contributing area thresholds: steep slopes can support channels with smaller contributing areas, while gentler slopes may require larger contributing areas to produce enough erosive power to incise and maintain a channel. The result is a network with a particular density and branching structure. That structure matters for hydrographs because channels are fast pathways. The moment water enters a channel, travel times shrink dramatically. A basin with a dense channel network, or with extensive artificial drainage, tends to respond quickly; a basin where much water must travel as slow subsurface flow responds more slowly.

This brings the discussion into the human world, where the "pipe" metaphor becomes dangerously accurate because we build pipes. Urban storm drains, straightened channels, and tiled agricultural fields all increase connectivity and reduce storage. They convert a catchment into something more like plumbing: less buffering, more flashiness. The hydrograph sharpens: shorter lag time between rain and peak flow, higher peak discharge, faster recession. The water that once soaked into soil or lingered in wetlands is delivered rapidly to channels and then downstream.

A concrete example is the transformation of small watersheds around expanding cities. Before development, a summer thunderstorm might produce a modest rise in a creek, because much rainfall infiltrated into soil, and the stream's response was damped. After development, the same storm can produce a rapid surge. Pavement and roofs are not just impermeable; they are also smooth collectors that funnel water quickly to drains. The drains connect to channels. The channels convey water downstream with little delay. In many places, the most vivid sign of this change is not the average river level but the suddenness with which it can rise, turning a benign channel into a hazard.

The history of hydrology contains a vivid anecdote about learning to read such pulses. In the 1850s, the Irish engineer Thomas Mulvany developed a method to estimate flood peaks for designing drainage works, relating peak discharge to rainfall intensity, catchment area, and a runoff coefficient. His ideas helped formalize the rational method, still taught in basic engineering hydrology. The broader point is not the particular formula but the historical pressure behind it: as people drained lands, built bridges, and expanded cities, they needed ways to anticipate the pulse. The demand for prediction made discharge a central concept and pushed hydrology toward quantification.

Yet discharge is not only about floods. It is about life in rivers. Many aquatic organisms time their reproduction, migration, and feeding to flow regimes. A spring snowmelt pulse can cue fish to move upstream; a summer low-flow period can concentrate nutrients and raise temperatures; a sequence of moderate floods can maintain side channels and wetlands that serve as nurseries. Alter the pulse with dams and diversions, and you alter habitats. A river's *breathing* is not poetic license; it is an ecological rhythm.

Seasonality adds another layer. In snow-dominated basins, winter precipitation is stored and released as a broad spring pulse, often with a predictable recession into summer baseflow. In monsoonal climates, the pulse may be concentrated into a few months with high variability from year to year. In maritime climates, frequent storms can produce many pulses superimposed on a high background. In each case, the catchment's internal stores control how much of the atmospheric input is translated into discharge and on what timescale.

Hydrologists sometimes distinguish between *quickflow* and *baseflow*.

Quickflow is the rapid component of runoff that produces the steep rising limb of a storm hydrograph, often associated with near-surface flow paths, saturated areas, and direct precipitation onto channels. Baseflow is the slower component supplied by groundwater and deeper storage, sustaining streams between storms. The distinction is conceptually helpful, but in reality these components blend. A catchment's pulse is a composite signal, made from multiple reservoirs with different response times. During a wet season, the reservoirs become more connected; during a dry season, connections break and the stream relies increasingly on slow release from deeper stores.

One of the most emotionally resonant moments in a river's pulse is the first storm after a long dry spell. In many places, that first storm does not produce a flood; instead, it produces smell and color. Streets and soils release accumulated dust and oils; streams darken with the first flush of organic matter; the water may carry a sharp, earthy scent. Hydrologically, the basin is recharging. Water is filling the unsaturated zone, raising groundwater, wetting channel beds that may have been dry. Only after repeated storms does the basin begin to respond more directly, with higher peaks for the same rainfall. This is the catchment learning to breathe again, opening pathways that had collapsed under dryness.

The inverse moment is late summer in a temperate basin: the channel shrinks, exposing gravel bars and algae-stained stones. To a casual observer, the river looks "small," but from the perspective of the network it is in a different operating mode. Much of the basin is disconnected. Small tributaries may not flow at all. The river is being sustained by a thin thread of groundwater discharge, sometimes called the *groundwater subsidy*. In a warming climate, this season can lengthen, and the helpful buffering of groundwater can be strained. Here the pulse becomes not only a response to storms but a measure of vulnerability.

If all this sounds complicated, there is a comforting simplicity at the core: conservation of mass. Water does not vanish inside a catchment; it is stored, delayed, transformed in phase, or exported. Discharge is the part of that accounting we can see most directly. It is the signature of how a landscape transforms atmospheric delivery into gravitational flow.

Standing beside a river during a rising stage can be unsettling because it reveals how quickly a basin can change state. The water climbs the bank, submerges roots, begins to lift twigs and leaves, and then whole branches. The current becomes louder, less conversational and more insistent. What you are hearing is not just water moving; you are hearing connectivity increase. Hillslopes that were quietly absorbing are now contributing. Hollows that were damp are now saturated. Small channels that were trickling are now conveying. The catchment is, for a time, more integrated as a machine.

Then the rain stops. The river begins to fall, not instantly but with a characteristic recession. That falling limb is the basin's exhale. Water drains from near-surface stores, then from deeper ones. The channel clears, bars re-emerge, and the sound softens. In the stones and mud left behind are the event's notes: a new line of debris marking the peak, a fresh layer of silt on a floodplain, a scoured patch of gravel where fish may later spawn. The pulse has passed, but its traces remain, a reminder that a river's shape and power are expressed not in a constant flow but in the living variability of discharge.

Chapter 2

The Tree of Water

Rain falling just inches apart can end up in oceans separated by thousands of miles, guided by the silent logic of the landscape's spine. By viewing rivers not as chaotic scribbles but as mathematical trees, we uncover the strict hierarchy that governs how continents shed their weight in water.

2.1 Drawing Lines in the Dirt

A landscape looks continuous. Hills fade into valleys; valleys widen into plains; the whole surface feels like a single, rumpled sheet. Yet the moment rain begins to fall, that sheet is quietly partitioned. Water does not simply go "downhill" in the vague way we say at the dinner table. It makes a choice at every bump, pebble, hoofprint, and root-thickened crease. Those choices are not arbitrary: together they sort the land into a mosaic of territories, each with its own exit route. A *drainage basin* (often called a watershed or catchment) is one of those territories: the area of land where all precipitation that becomes runoff ultimately converges toward the same outlet—a particular stream, a lake, an estuary, or a point on a coastline.

Drainage basins are the fundamental unit of geomorphology not because geomorphologists enjoy drawing boundaries on maps, but because basins turn the messy business of weather into something the

Earth's surface can negotiate with. Rain arrives scattered in space and time, carried by storms that care nothing for the shape of a valley. Topography converts that scattered input into organized flow: sheet to rill, rill to channel, channel to trunk river. In other words, a basin is a piece of landscape that acts like a funnel. Once you name the funnel, you can begin asking what it does: how quickly it moves water, how much sediment it supplies, how it responds to droughts and deluges, how it grows or shrinks as divides migrate, and why one catchment becomes a gentle agricultural valley while its neighbor cuts a steep gorge.

The edge of a drainage basin is a *divide*. In the simplest picture, a divide is a ridge: step to one side and a raindrop will eventually reach one river; step to the other and it will reach another. That sounds almost trivial until you confront what the word "eventually" hides. A divide can separate not only two nearby streams but two distant destinies: different deltas, different seas, different oceans. It can also separate different hazards and different livelihoods: one town's drinking-water source from another's; one valley's flood risk from another's; one reservoir's sediment supply from the next reservoir over. A divide is a line in the dirt that the atmosphere routinely tests, drop by drop.

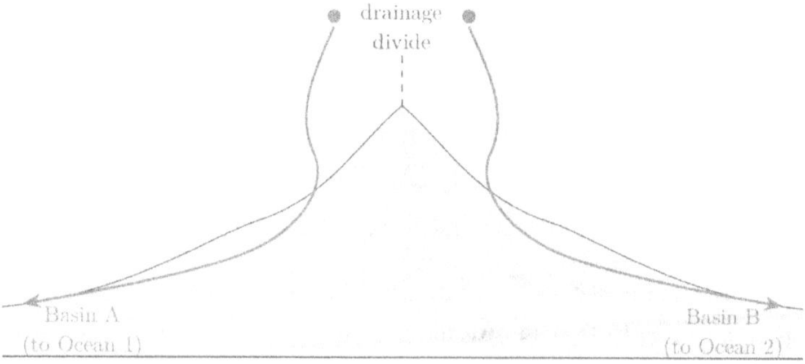

The cartoon simplicity of a ridge and two raindrops is useful precisely because real divides are rarely so clean. On a detailed topographic map, the boundary of a basin wriggles along high points: sometimes a sharp crest, sometimes a subtle saddle between two hummocks, sometimes a surprisingly flat stretch where the "high point" is barely higher than a soccer ball. The rule is not that a divide must look dramatic. The rule is topological: the divide is the set of points where flow paths separate. If you released a tiny, frictionless bead of water at each point and let gravity choose the path of steepest descent, the bead on one side would be captured by one outlet, the bead on the other side by another.

That little mental bead introduces the first essential idea: a drainage basin is defined by *connectivity*, not by appearance. To know which basin a point belongs to, you do not need to know the color of the rock or the type of vegetation. You need to know where water from that point can go. Basins are therefore naturally suited to what might sound like an abstract branch of mathematics, topology, because topology cares about connections, not exact measurements. A basin is the region of the landscape connected by downhill routes to the same exit. The fine details of those routes will change with soil moisture, vegetation, and microtopography, but the broad partition often persists for long enough to shape valleys, transport sediment, and carve the map's familiar dendritic patterns.

The second essential idea is energetic. A basin is not simply an area; it is an engine that converts potential energy into motion and work. Rain starts with gravitational potential energy because it lands at some elevation above the outlet. As water moves downhill it spends that energy, and in spending it, it entrains grains, abrades bedrock, undercuts banks, and sorts sediment into bars and floodplains. The basin boundary matters because it defines the accounting ledger for that energy. All else equal, a larger basin collects more water, and that water has more opportunity to concentrate into channels capable of moving larger sediment. This is why the same storm can leave one valley with a gentle rise in streamflow and another with a wall of water: not simply because of rainfall totals but because of the way each basin gathers and routes that rainfall.

There is a famous American version of the raindrop story that makes the notion of a divide feel less like a technicality and more like a continental drama. High in the Rocky Mountains, the Continental Divide separates waters headed toward different oceans. A drop falling on one side may journey through the Columbia River system toward the Pacific; on another, it may travel through the Missouri and Mississippi toward the Gulf of Mexico and the Atlantic. The actual routes are complicated, of course, and the exact location of a divide can be surprisingly nuanced, but the rhetorical force is real: a step can change an ocean. That kind of boundary is why divides appear not only in scientific papers but in national park brochures and school lessons. The concept is a way of thinking about land as a set of choices made by gravity.

Yet, if divides sound permanent, landscapes quickly teach humility. The boundary of a basin is not an inked line; it is a moving compromise between competing erosion systems. Imagine two neighboring valleys chewing backward into a plateau. Each valley's headwaters erode their way upslope through processes with unglamorous names—rilling, gullying, frost cracking, landsliding—and one valley may eventually cut across the other's territory. When that happens, the divide jumps. Water that used to belong to one basin becomes the property of another. Geomorphologists call this *stream capture* or *river piracy*, a term that sounds like folklore but describes a real rewiring of the drainage network. It leaves behind telltale signatures: an oddly sharp elbow in a river, a dry notch (a wind gap) where water once flowed, a mismatched valley too large for the stream that now occupies it. The important point is that the basin is still the right unit, even though its boundary can migrate; in fact, because capture changes which slopes feed which channels, basins are the natural way to describe what has been stolen and what has been gained.

To see why basins became so central to the science, it helps to visit the mid-twentieth century, when geomorphology began to quantify what earlier observers had mostly described. Robert E. Horton, working in the 1930s and 1940s, treated a drainage basin as something you could measure systematically: the total length of streams, the number of tributaries of different sizes, the spacing of channels across the surface. He was not the first person to look at maps and notice branching patterns, but he helped make those patterns countable. Once you count them, you can compare one basin to another. A basin in humid Appalachia and a basin in semi-arid New Mexico may look different to the eye, yet their networks can be summarized by ratios and scaling relationships. That shift—from picturesque description to quantitative morphology—gave geomorphology a durable set of tools, and it did so by insisting that the basin is the natural container for those tools. A single river reach is too local; an entire continent is too heterogeneous. A basin is just right: bounded, coherent, and driven by a shared outlet.

Arthur Strahler sharpened the same idea in the 1950s. One reason basins are so useful is that they allow a separation between quantities that depend on size and quantities that do not. The area of a basin obviously depends on how big it is, so does the length of its main channel. But certain *dimensionless* descriptors—ratios of counts or lengths across stream "orders," for instance—can be meaningfully compared across basins of different size. That was a quiet revolution in thinking. It meant that a small basin could be treated as a miniature of a large one, not in every detail, but in the sense that both might obey similar organizing rules. This idea of geometric similarity is never exact in nature, but it is powerful enough to guide hypothesis and measurement.

All of this might still feel like the kind of abstraction that lives comfortably on paper and uncomfortably in mud. So take the idea back outside, to a place almost anyone can encounter it: the edge of a parking lot during a heavy rain. The asphalt is a miniature watershed. Water beads on the surface, then begins to slide, guided by tiny tilts you would not notice in dry weather. Painted lines, curb cuts, and the faint sag where a truck has repeatedly parked become topographic controls. Soon the flow concentrates into a thin thread that traces a route toward a storm drain. If you scattered a handful of rice grains on the wet pavement, you would watch them migrate in converging paths, collecting where the flow concentrates. A curb is a divide; a low point at the end of the lot is an outlet; the storm drain is a kind of engineered river mouth. The same logic scales up to a hillside and then to a mountain range. What changes with scale is not the basic rule of connection but the complexity of the pathways and the amount of work the water can do.

That scaling is easier to appreciate if the basin is treated not as a static area but as a *system* with inputs, storage, and outputs. The input is precipitation (and in cold climates, stored snow and ice that later melt). Some of that water infiltrates into soil and rock, moving slowly as groundwater; some evaporates or is transpired by plants; some becomes surface runoff. The output is discharge at the basin's outlet: the flow you can measure with a gauge. A hydrograph—a plot of discharge versus time—is a kind of signature of how a basin behaves. Two basins receiving the same rainfall can produce dramatically different hydrographs. A steep basin with thin soils and sparse vegetation may respond quickly, with a sharp peak. A basin with deep soils, wetlands, or permeable geology may respond slowly, spreading the same input over days or weeks. Even without numbers, people experience this difference: the creek that becomes dangerous an hour after a thunderstorm versus the river that rises calmly overnight.

The basin perspective also clarifies why geomorphology is inseparable from sediment. Water is the mover, but sediment is the cargo that changes the channel that carries it. Within a basin, different parts of the landscape supply different kinds of sediment. A landslide delivers boulders and shattered blocks; a soil-mantled slope delivers silt and sand; a glaciated headwater may deliver finely ground flour-like sediment. Those sources are not distributed uniformly. They are organized by slope, by vegetation, by geology, by the history of fire and storms. A basin boundary therefore encloses not only a set of flow paths but a set of sediment pathways. When engineers worry about a reservoir filling with silt, they are really worrying about the reservoir's upstream basin: every eroding bank and every gullied hillside within that boundary is, in a long accounting, contributing to the deposit at the dam.

If the basin is the unit of accounting, the divide is the auditing line. But divides can be deceptive auditors. Some basins have no outlet to the sea. These are *endorheic* basins, closed depressions where water collects and leaves mainly by evaporation or seepage, often leaving salts behind. The Great Salt Lake sits within such a system, and many desert basins around the world share the same logic. In these landscapes, water still flows downhill and organizes channels, but the destination is internal. The basin becomes even more visibly a container, because the outlet is not a river mouth but a lake, a salt flat, or a playa that records wet years and dry years like rings in a tree stump. Thinking in basins prevents a common misunderstanding: not every river is trying to reach the ocean. Some are trying, unsuccessfully, to escape a bowl.

Divides also blur in flat terrain, where the "highest ground" is hard to identify and water can spread laterally. In peatlands or floodplain complexes, small changes in vegetation, beaver dams, or the growth of natural levees can redirect flow. During floods, a river may temporarily connect to neighboring basins through overbank flow, creating short-lived pathways that a dry-season map would miss. At the largest scales, nature sometimes breaks the tree-like assumption entirely when channels split downstream into distributaries, as in deltas. That is not a failure of basin thinking; it is a reminder to be precise about what is being bounded. A drainage basin is the area contributing water to a particular outlet, but in a delta there may be multiple outlets whose shares change over time. The boundary upstream may still be crisp, while the boundary downstream becomes a shifting lacework.

Despite these complications, the basin remains the simplest powerful idea for organizing the messy intersection of climate, gravity, rock, and life. It links the unremarkable act of a raindrop hitting soil to the slow carving of valleys. It explains why two neighboring hillsides can live in different hydraulic worlds. And it offers a moral, in the plain sense of a practical lesson: the lines that matter most on a landscape are often invisible until the rain arrives. Then, all at once, the land declares its allegiances, and every slope becomes an argument about where the water belongs.

2.2 The Calculus of Branches

A drainage basin is a promise: every drop has an address, and every address has a route to an outlet. Yet even a small basin can look bewildering when it is drawn faithfully. The map fills with hairline tributaries, some seasonal, some permanent, some no more than a wet crease in clay. Seen from above, the pattern resembles a leaf's veins or the branching of a bronchial tree, a geometry that feels orderly without being obviously designed. The mind quickly reaches for metaphors, but engineers and geomorphologists need arithmetic. They need a way to say, without hand-waving, whether one basin is more branched than another, whether a channel is a "main river" or merely a side groove, and how network complexity builds as water concentrates.

That need produced one of the most practical pieces of river logic ever invented: *stream order*. The version most widely used today is associated with Arthur Strahler, who refined earlier ideas in 1957. It is almost childishly simple, which is exactly why it stuck. You start at the tips of the network, the smallest unbranched channels, and label them order 1. When two channels of the same order meet, the order increases by one. When channels of different orders meet, the downstream segment keeps the higher order. The rule is a kind of bookkeeping for branching: it tallies genuine mergers of equals, while treating the addition of a small tributary to a large river as a detail that does not change the river's rank.

Behind that simplicity is a subtle claim about how river networks work. They are not just collections of lines; they are hierarchies. Not every junction is equal. Some are decisive, changing the scale of flow and the width of the valley; others are small side additions, easily lost in the noise of storms and seasons. Strahler's ordering system turns that qualitative sense of hierarchy into something that can be counted, compared, and related to other features of the basin.

The ordering rules feel more intuitive if the language is kept plain. A first-order stream is a fingertip: water that has found a channel but has not yet met another channel. Two fingertips join to make a second-order stream. Two second-order streams join to make a third. And so on. Crucially, a first-order joining a second-order does *not* make a third-order. That junction may be dramatic to an angler who knows the cold tributary that brings trout into the main stem, but in the network's hierarchy it is not a step-change. This "1+1=2" and "2+1=2" logic is what makes Strahler order feel like arithmetic rather than naming.

It is also what makes the system resistant to the endless ambiguity of natural channels. Rivers do not come with labels, and even the act of deciding what counts as a channel is trickier than it sounds. Do you include ephemeral rills that appear only in heavy rain? What about gullies that carry water a few weeks each year? Different maps, different field crews, and different climates will draw different answers. Stream order does not solve that problem, but it gives a stable way to proceed once you have made a choice. Begin at the tips you are willing to call channels, apply the rules consistently, and the hierarchy follows.

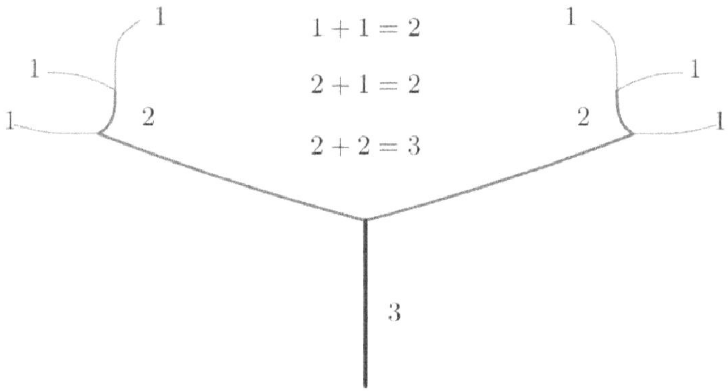

Once the labels are on the network, a basin that looked like scribble begins to reveal a calm internal logic. Stream order makes it possible to say things that are otherwise slippery. You can talk about how many first-order streams feed the system, how quickly order increases downstream, and what the maximum order is. That maximum, the order of the trunk at the outlet, is a compact way to express overall network size and complexity. A tiny headwater basin may only reach third order; a continental-scale system like the Mississippi reaches far higher.

There is a satisfying historical irony here. Rivers were once described with names that emphasized personality—wild, braided, meandering, youthful, mature—as if they had temperaments. Horton and Strahler helped replace temperament with tally marks. Horton, writing in 1945, had already set out a stream ordering method and a suite of ratios to capture network structure. Strahler's refinement, now taught in introductory classes, was in part an attempt to make the ordering unambiguous and robust for analysis. The goal was not to flatten nature into numbers but to give numbers a chance to speak about nature in a disciplined way.

Those numbers do speak, and not only in the abstract. One of Horton's most enduring observations is that stream networks tend to show regularities when you move from one order to the next. The number of streams usually decreases systematically with increasing order; lengths and drainage areas tend to increase. These are sometimes called Horton's "laws," not because they are as rigid as physics but because they recur across basins often enough to be useful. Strahler emphasized that many of the most informative descriptors are *dimensionless ratios*: for example, the *bifurcation ratio*, which compares the number of streams of one order to the number of the next higher order. Ratios like this have a social charm: they allow comparison between basins that differ greatly in size. A small basin and a large basin can share a similar bifurcation ratio, suggesting a kind of family resemblance in their branching architecture even if their maps are drawn at different scales.

Why should a branching network display such regularities at all? A river network is not built by a single blueprint; it is assembled incrementally by storms, by erosion, by vegetation changes, by landslides, by the slow creep of divides. Yet the outcomes often look statistically similar. This is one of the deepest pleasures of geomorphology: the sense that order emerges from countless local accidents.

To make that emergence feel less mystical, consider what stream order is really doing. It is translating geometry into hierarchy. At each junction, the ordering rule asks: is this merger the union of comparable tributaries, or is it the addition of a small side channel into a larger one? The first case changes the rank; the second does not. That distinction lines up with hydraulics in a rough but meaningful way. When two similar streams meet, the downstream channel often widens noticeably, the valley opens, and the floodplain changes character. When a small tributary enters a large river, the large river may barely react in its overall dimensions, even if the confluence is locally turbulent. Strahler order is not a measurement of discharge, but it is a proxy for the nested structure that discharge tends to follow.

There is a further, more philosophical advantage. The ordering system is a way of treating the basin as a tree. Trees have a root (the outlet), branches, and tips. A river network is not a living tree, but at the scale of an ordinary drainage basin it behaves like one: flow converges, not diverges. That tree-likeness allows the tools of network thinking to enter geomorphology. It makes it sensible to speak of upstream "ancestors" of a channel and downstream "descendants," of path lengths, of branching probabilities. Even without mathematics, the tree image changes how the mind handles the complexity. Instead of seeing an indecipherable maze, you see a hierarchy of mergers.

Still, any claim to order earns suspicion when it meets fieldwork. Take a concrete case: a familiar mountain watershed that supplies a city's water. In the Colorado Front Range, for instance, dozens of small headwater channels drain snowmelt into a few larger forks, which then feed reservoirs and diversions serving communities along the foothills and plains. The water managers care about peak flows and storage, but they also care about sediment and debris flows after wildfires. Stream order helps organize that concern. The first-order channels are often steep, closely coupled to hillslopes, and sensitive to disturbances like tree loss. After a fire, intense rainfall can strip ash and soil into these small channels, which then deliver a pulse downstream. Whether that pulse becomes a manageable increase in turbidity or a damaging debris flow depends on many local factors, but the hierarchy matters: disturbances initiate in the small tips and compound as they merge. A map of orders is not a forecast, yet it is a way to think clearly about where problems begin and how they scale.

Order also clarifies a point that casual river-talk often muddles: "the river" is not one thing. It is a sequence of linked environments. Headwater channels, often first and second order, can be cold, shaded, and full of large wood. Their flow may wink out in dry seasons, yet they are crucial for ecology and for sediment supply. Mid-order channels collect those headwaters into something more persistent, with alternating riffles and pools, with banks that begin to store fine sediment. Higher-order trunk rivers carry the accumulated water and sediment, constrained by floodplains and valley geometry. Stream order offers a simple language for that progression without needing to specify exact widths or discharges.

Strahler's 1957 paper is also remembered for insisting on a disciplined separation between measurements that scale with basin size and ratios that may reflect deeper structure. That insistence remains relevant because it guards against a common error: mistaking bigness for complexity. A large basin has more channels simply because it has more area. Stream order helps reveal whether it has more *hierarchical levels* or merely more repetitions of small parts. Two basins can have similar maximum order but very different total area because one has long, widely spaced tributaries and the other has dense, closely spaced ones. That difference matters for flood response, sediment yield, and habitat connectivity. Without a hierarchy measure, it is easy to miss.

Hierarchy also provides a bridge to one of the most celebrated scaling relationships in fluvial geomorphology: *Hack's law*. Working for the U.S. Geological Survey in the 1950s, J. T. Hack examined stream profiles and basin geometry and helped establish that the length of the main channel, L, tends to scale with drainage area, A, as a power law,

$$L \sim A^h,$$

with the exponent h often near 0.6, though it varies. A power law is a compact way of saying that doubling the area does not simply double the main channel length; the relationship bends. The striking thing is not the precise exponent—geomorphologists still debate why it takes the values it does, and modern large-sample studies emphasize that basin shape and geometry can influence the fitted parameters in ways not easily pinned on climate or rock type. The striking thing is that a basin's geometry is constrained enough to produce such a relation at all.

Stream order and Hack's law share a family resemblance. Both assume that river networks have regularities across scale. Stream order does it with integers and junction rules; Hack's law does it with a scaling curve. Neither is a law in the sense of Newton, yet both work often enough to organize thought and to guide measurement. Together they hint at a deeper claim: river networks are not just accidents of where rain happened to fall. They are self-organizing structures shaped by gravity, erosion, and the tendency of flow to concentrate along paths of least resistance.

That phrase—paths of least resistance—opens the door to a more modern, and more contentious, way of thinking about exponents and ratios. Some researchers have argued that certain scaling behaviors in river networks resemble those in statistical physics, such as percolation, where flows find connected pathways through heterogeneous materials. Allen Hunt (2016), for example, used percolation-based reasoning to suggest that the spread in observed Hack exponents might reflect differences in substrate heterogeneity and tortuosity rather than being a simple fingerprint of climate. One does not need to accept any single theory to appreciate the intellectual shift: the basin is being treated less like a landscape curiosity and more like a network whose geometry emerges from general principles.

Yet the practical appeal of Strahler order remains its modesty. It asks for no grand theory. It simply offers a rule that can be applied to a map, a digital elevation model, or even a careful walk along a creek with a notebook. You can take a basin you have never seen, label its tips as 1, work downstream, and end up with a compact description of its branching hierarchy. That description can then be linked to questions people actually care about: how quickly flood peaks propagate, how sediment pulses move, how habitats connect, where pollutants introduced in small tributaries might concentrate, and how a basin might respond if its headwaters are logged, burned, urbanized, or simply altered by a changing climate.

There is also an aesthetic payoff. Once you learn to see order, a river map becomes less like a tangle and more like a statement. The smallest channels are not failures to reach riverhood; they are the network's sensory organs, sampling rainfall across the basin. The mid-order tributaries are the integrators, gathering those samples into coherent flow. The highest-order trunk is the messenger, carrying the basin's combined signal downstream. When the storm ends and the hydrograph falls, that hierarchy remains etched into the land, a quiet record of how the basin has learned to collect water and turn it into work.

2.3 When Waters Collide

A river network looks calm on a map because ink cannot show hesitation. Two lines meet, one line continues, and the page pretends the merger is instantaneous. Stand at a real confluence and the pretense dissolves. Water arriving from two valleys is never identical. One branch may be colder, shaded by spruce and fed by snowmelt; the other may be warmer, sunlit, and running off clay slopes. One may carry fine silt that turns it the color of tea with milk; the other may be clear enough to reveal pebbles on the bed. They collide at a junction that seems as ordinary as a street intersection, yet it is one of the most active places in the whole basin: a site where momentum, sediment, and chemistry must negotiate a shared channel.

The word *confluence* sounds gentle, a flowing-together. The physics is not always gentle. Each tributary arrives with a certain speed, depth, and direction. Those traits are not cosmetic; they are the river's way of expressing energy. When the two flows meet, energy has to be redistributed. Some becomes turbulence, the chaotic swirls that grab sediment off the bed and mix different waters. Some is spent reshaping the channel itself—eroding a bank here, building a bar there—as the merged flow tries to find a stable path. Confluences are where a drainage network does its integrating, turning a tree of many inputs into a single output, and the integration has consequences.

A useful way to see what is happening is to think in terms of *momentum*, which is just mass times velocity. Water is heavy, and moving water resists being redirected. If two tributaries meet at an acute angle, the larger one often bends the smaller into line, sweeping it along the bank in a narrow strip. If they meet more symmetrically, or with comparable discharges, neither can dominate immediately. The joined flow tends to fan out, slow down, and then gather itself again downstream. In that gathering, the channel bed and banks become a kind of referee. The channel deepens in some places and shoals in others until the flow has a shape that suits it.

One of the most striking features at many confluences is the *mixing interface* between the two waters. People sometimes call it a "line," but it is better imagined as a zone: a shear layer where two streams slide alongside each other at different speeds. Shear makes turbulence, and turbulence makes mixing, but mixing is not instantaneous. Over a short distance you can often see a two-toned river, a visible record of different sediment loads. Even without color, the difference can be measured: a tributary rich in dissolved ions from limestone may have a different electrical conductivity than a tributary draining granite; a tributary warmed by a wide, shallow floodplain may carry a distinct temperature signal; a tributary with algae may carry different nutrients. The confluence becomes a place where a basin's varied geology and ecology meet in the same cross-section.

There is a famous example that has become a kind of informal laboratory for the public imagination: the meeting of the Rio Negro and the Solimões near Manaus in Brazil, often referred to as the "Meeting of Waters." The dark, tea-colored Rio Negro meets the lighter, sediment-laden Solimões, and for kilometers they run alongside each other with a visible boundary. Tour boats idle on the seam so that visitors can watch the stubbornness of partial mixing. The spectacle is not magic; it is a consequence of differences in temperature, density, and suspended sediment, combined with the scale of the flows. But it dramatizes a general point: at a confluence, the river's unity is something that must be achieved, not assumed.

The mechanics of that achievement are intimately tied to sediment. A tributary does not arrive empty-handed. It brings grains in suspension and grains rolling or hopping along the bed. It also brings the memory of its last storm: the coarse gravel flushed out by a cloudburst, the fine clay carried for days after an upstream bank failure. When two sediment-laden flows meet, the capacity of the merged flow to carry sediment changes. Capacity is not a moral concept; it is simply the amount of material a flow can keep moving without depositing it, given its speed, depth, and turbulence. If the joined flow slows down, sediment drops out. If it accelerates or becomes more turbulent, it can pick up more.

This is why confluences are often associated with bars and islands. The combined river may not have enough energy to carry all the sediment supplied by both tributaries, especially during falling flows when the water level drops but the bed still holds yesterday's load. A wedge of sediment can form at the junction, forcing the merged flow to split around it and scour one side more deeply. Over time, the confluence develops a characteristic geometry: a deep *scour hole* near the point where fast currents strike the bed, flanked by bars where flow decelerates and deposits sediment. The scour hole can be unexpectedly deep, sometimes the deepest spot for many kilometers, because it is created by a focused combination of flow convergence and turbulence.

Confluences also create ecological edges. Many fish use them as meeting places because the mixing zone can concentrate food and create hydraulic shelter. In a clear tributary entering a turbid main stem, predators may patrol the seam, exploiting the contrast in visibility. Temperature differences can matter too: a cold tributary can offer refuge during hot spells, and its plume may persist downstream as a ribbon of cooler water near one bank. In this way the geometry of branching becomes a geography of habitats, and the arithmetic of stream order acquires a biological texture.

The human consequences of confluences are older than hydrology as a science. Cities often grow near junctions because junctions are natural nodes for travel and trade. Pittsburgh sits where the Allegheny and Monongahela join to form the Ohio River, a location that mattered long before the steel era because it is a place where the landscape offers a decision: go upstream into one basin or the other, or go downstream into the greater river corridor. Cairo, Illinois, sits near the meeting of the Ohio and Mississippi, a confluence that has repeatedly forced the city to confront floods and levees. In South Asia, the Ganges and Yamuna meet at Prayagraj (Allahabad), a confluence that is both hydrologic and sacred, drawing millions to a place where waters of different origins are believed to retain their identities even as they join. When people gather at a confluence, they are responding to the same thing a geomorphologist sees: a junction is an amplifier of connectivity.

The science of stream ordering fits naturally into this picture because it treats confluences as the decisive events in the network. Horton's and Strahler's rules are, at heart, rules about junctions. A confluence between two streams of equal order is treated as a structural step-change. That bookkeeping is more than pedantry. It encodes a real asymmetry between equal mergers and unequal ones. Equal mergers are rarer but consequential; unequal mergers are common but often locally confined in their effects. A small tributary can deliver a striking pulse of mud that stains a river for kilometers, yet the river's overall width and slope may barely shift. A merger of equals can change the valley itself.

The details depend on discharge ratio (how much water each tributary contributes), sediment ratio, and junction angle. Imagine two tributaries approaching the confluence with similar discharges but different sediment loads. The clearer stream may be hungry for sediment—its flow could carry more than it currently has—and when it meets a sediment-rich tributary, it may increase its abrasive power. The merged flow can then scour the bed downstream, sometimes undermining banks that had been stable for decades. Conversely, if both tributaries are already heavily loaded with sediment and the confluence forces a reduction in velocity, deposition can dominate, building bars that redirect the main flow toward a bank and trigger lateral erosion. The same confluence can alternate between these behaviors across seasons: scouring during floods, depositing during recessions.

A confluence also introduces a kind of hydraulic politics: which tributary gets the best alignment downstream? A tributary entering at a steep angle can be pushed against the outer bank, where faster flow tends to erode. Over time, that tributary's momentum can help steer the combined channel, increasing curvature downstream. In meandering rivers, this steering can affect where bends migrate and where cutoffs might eventually occur. In gravel-bed rivers, where the bed is mobile and the channel often splits around bars, confluence dynamics can decide which branch becomes dominant after a flood. Confluences are therefore not just points; they can imprint patterns for kilometers.

If confluences are where waters meet, there is another, quieter kind of collision that happens at basin boundaries: the collision between two drainage systems competing for the same high ground. A divide is not a wall; it is a ridge being eroded from both sides. Given enough time, one side can win, not by pushing the other back directly, but by eroding headward faster and intercepting the other's tributaries. When that happens, the confluence of waters is replaced by a rerouting of waters, and a basin is rewritten.

This process, stream capture, is easiest to picture with a simple thought experiment. Two neighboring valleys cut into a plateau. One valley has a steeper gradient to base level (perhaps it drains to a lower river or a nearer coastline), so its streams have more energy per unit length. They erode their headwaters more aggressively, nibbling upslope. Eventually a headwater channel crosses the divide and intercepts flow that once fed the other basin. The captured stream is now forced to turn, often abruptly, toward its new outlet. Downstream, the capturing basin gains discharge and sediment supply; the victim basin loses them. The map changes, but so does the erosional future. Increased flow can deepen the captor's channels, further enhancing its advantage. The victim may be left with an underfit stream in an oversized valley, like a thin thread in a trench.

As summarized in Wikipedia's entry on stream capture, geomorphologists look for clues such as elbows of capture and wind gaps—dry valleys that once carried water but no longer do. These are not mere curiosities. In regions where water supply is sensitive, capture can alter which communities sit upstream of which, and in mountain belts it can reorganize sediment delivery to foreland plains. Capture is confluence's shadow image: instead of two flows joining, one flow is stolen before it can join, and the network's supposed permanence is revealed as provisional.

There is a further complication, one that seems at first to violate the very idea of a river network as a tree. Trees converge; rivers, most of the time, converge too. But there are places where a single channel splits into two and both branches persist. This is *bifurcation*, often associated with deltas, distributary systems, and sometimes braided channels. As summarized in Wikipedia's entry on river bifurcation, bifurcations can be temporary, forming around sediment islands that shift from flood to flood, or they can be stabilized by vegetation, levees, or bedrock controls. In a delta, a bifurcation is the mechanism by which a river spreads its sediment and builds land, dividing its load among multiple mouths.

From the perspective of confluences, bifurcations matter because they change the meaning of "downstream." In a purely tree-like network, every point has one path to the outlet. In a bifurcating system, there can be two, and the share of flow sent down each branch can change with discharge. That turns the network into something more like a braided conversation than a single chain of command. Even inland, bifurcations occur where channels are perched on alluvium and small changes in bed elevation can send water one way or another. In some rare cases, a bifurcation is so persistent that it links major basins, allowing water to drain toward two different outlets depending on season or long-term change.

These departures from perfect tree-ness do not make stream ordering obsolete; they make it more interesting. A confluence is a moment of integration; a bifurcation is a moment of choice. Both are places where the network's topology becomes visible in motion. They are also places where human interventions are most tempting and most risky. Engineers straighten channels, armor banks, dredge bars, and build levees, often with the goal of stabilizing junctions for navigation or flood control. But a confluence wants to adjust. Constrain it too tightly and the river will spend its energy elsewhere, perhaps by deepening a scour hole that undermines a structure or by eroding an unprotected bank downstream. Rivers are not vindictive, but they are persistent, and confluences are where persistence concentrates.

A concrete example of that persistence can be found in the history of river engineering on large alluvial rivers, where confluences and their bars posed hazards for navigation. On the Mississippi system in the nineteenth and early twentieth centuries, shifting sandbars at tributary mouths could redirect the main current, shoal shipping lanes, and alter bank erosion patterns near settlements. The response was often to "train" the river with structures that narrowed and guided the flow. Some of these efforts succeeded locally while triggering surprises elsewhere, a reminder that a confluence is not a point problem but a network problem: adjust one junction and you alter energy and sediment pathways that extend upstream and downstream.

At the scale of a single storm, confluences can behave like instruments reading different parts of the basin. One tributary may peak early because its slopes are steep and its soils thin; another may peak later because its basin stores water in wetlands or deep soils. When those hydrographs meet, the timing matters. If peaks coincide, the downstream river can exceed flood thresholds dramatically. If they are offset, the flood wave may be stretched out, lower in peak but longer in duration. People living near junction towns often learn this timing intuitively. They know which tributary "comes up" first, which stays high, which turns muddy quickly. The confluence is where that local knowledge becomes legible as combined risk.

There is a final, almost philosophical aspect to confluences that is easy to miss if one stays at the level of physics and maps. A basin collects differences. Each tributary drains a particular set of slopes with a particular geology and vegetation. Those differences are not erased when waters meet; they are carried forward, sometimes for long distances, in chemistry, temperature, and sediment texture. A confluence is therefore a kind of archive. If the upstream basin experiences a landslide, the main stem carries the signal as a pulse of coarse sediment and elevated turbidity. If the upstream basin is mined, the signal may appear as dissolved metals or acidity. If the upstream basin is urbanized, the signal may arrive as warmer runoff, altered nutrients, and flashier hydrographs. The confluence is where signals superimpose, where a river becomes a sum of histories.

Stand at a junction long enough and you begin to see that the river is not one thing flowing past, but a continuing argument among sources. The seam between two waters may persist for minutes or for kilometers, but even when it disappears to the eye, the merger continues in other forms: eddies exchanging parcels of water, grains being sorted into bars, dissolved ions mixing molecule by molecule, microbes meeting new nutrients, fish sensing temperature gradients. The network's order is not only drawn in lines on a map; it is enacted at each confluence, where the basin's scattered experiences are forced into a single moving body, and where the landscape, in turn, is reshaped by the terms of that forced agreement.

Chapter 3

Nature's Hidden Geometry

To the casual observer, a river's path appears to be a triumph of randomness over terrain, a chaotic scribble across the map. This chapter peels back the landscape to reveal the rigorous statistical laws and fractal architectures that govern how water organizes itself, proving that the shape of a basin is determined less by local geology and more by universal mathematics.

3.1 The Hierarchy of Streams

Walk along almost any hillside after a hard rain and you can watch a river network being born in miniature. Water gathers first as a sheen, then as thin threads that hesitate, disappear into soil, reappear, and finally commit to tiny grooves. Some threads join and carve a slightly deeper line; others fade when the rain eases. From a distance it looks messy, improvised, even accidental. Yet when the same logic is scaled up to a whole landscape, the result is so regular that it can be counted, compared, and even predicted. The trick is to stop thinking of a river as a single line and start thinking of it as a family tree.

That shift in viewpoint was one of Robert Horton's gifts to the study of landscapes. Horton worked in the early twentieth century, at a moment when hydrology was becoming a quantitative science, pulled between field observation and the practical demands of flood forecasting, engineering, and water supply. He wanted numbers that could travel: measures that meant the same thing in a Catskills watershed and a desert wash, so that one basin could be compared to another without relying on a geologist's intuition. A stream network, he realized, is not merely a feature drawn on a map; it is an *organized* object. If you treat it as a branching structure, then order can be assigned to what initially looks like chaos.

The first step is disarmingly simple: identify the smallest channels at the tips of the network and call them the first. These are headwater streams, the outermost capillaries of the landscape. Some flow year-round; others appear only in wet seasons; some are so ephemeral that they may be visible only as a faint incision, a line of greener grass, or a chain of puddles after a storm. In Horton's view, they are not negligible scribbles but the fundamental units from which the rest is built.

Once the first-order pieces are named, the rest of the hierarchy follows a rule that feels more like arithmetic than geography. When two streams of the same order meet, they produce a stream of the next higher order. Two first-order streams join to form a second-order stream. Two second-order streams join to form a third-order stream, and so on. When streams of different order meet, the larger order persists downstream: a first-order stream joining a second-order stream does not "upgrade" it; it is absorbed as a side branch. This is the core of what later became widely known as the Horton—Strahler ordering system, after Arthur Strahler's mid-century refinements clarified its use and made it easier to apply consistently.

Placed early on a map, this idea immediately changes what the map *means*. Suddenly, every blue line is not just "a stream," but a member of a graded lineage. The question shifts from "where does the river go?" to "how is this branching arranged?" It becomes possible to say, in a well-defined way, how many first-order streams a basin has, how many second-order streams, how many third-order streams, and so forth. A basin is no longer an undifferentiated patch of terrain drained by a river; it is a nested structure that can be counted.

Horton–Strahler Stream Ordering (Schematic)

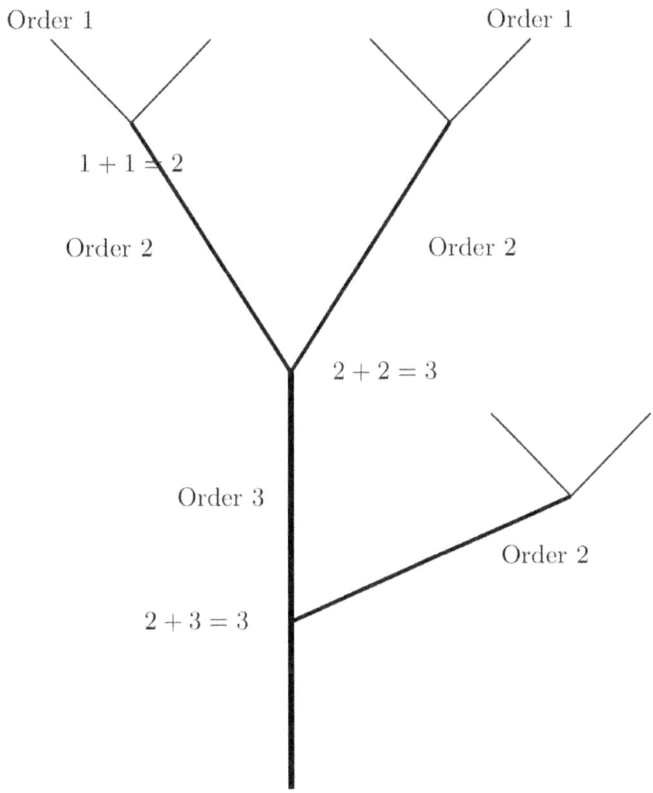

The diagram looks almost childlike, but it captures something profound: the network can be read as a set of local rules that build a global architecture. The "trunk" of a big river is not simply bigger because it happens to be lower on the map; it is bigger because it is the outcome of repeated, rule-governed mergers. This makes river networks resemble biological circulatory systems, lightning patterns, bronchial trees, and even the branching logic of some social networks. What matters first is not the material of the branch but the *topology*: which pieces connect to which.

To see why Horton's ordering matters, it helps to imagine a concrete watershed, even an ordinary one. Picture a hilly basin in the northeastern United States, maybe a few tens of square miles, drained by a creek that people fish in and that occasionally floods a road under spring snowmelt. On a detailed topographic map, the blue lines near the ridge crests are short, steep, and numerous. Those are mostly first-order streams. They are fed quickly by local rainfall and local groundwater; they respond fast and they dry fast. Slightly lower, where the valley sides have gathered enough contributing area to maintain a more persistent channel, the first-order threads join into a second-order stream: longer, deeper, more likely to carry water through the summer. Farther down, two second-order valleys meet at a confluence shaded by hemlocks, and from that point onward the stream is third-order. Even if you never compute a single ratio, the hierarchy is visible in lived experience: the headwaters are intimate and fickle, the mid-level channels are reliable, and the trunk is a different creature altogether.

Horton's classification gave geomorphology a way to quantify that intuition. Once a network is ordered, you can count how many streams there are of each orders. And this is where a new kind of regularity enters. Horton noticed that in many basins the number of streams declines in a roughly geometric way as order increases: there are many first-order streams, fewer second-order, fewer third-order, and so on, often by a nearly constant factor from one step to the next.

That factor is commonly called the *bifurcation ratio*. If a basin has, say, about four times as many first-order streams as second-order streams, and about four times as many second-order as third-order, then its bifurcation ratio is near 4. Real basins are never that perfectly tidy, but the striking point is that the ratio is often surprisingly stable within a basin and often comparable across very different regions.

At first glance, such a ratio can feel like numerology: why should landscapes care about tidy numbers? The answer is not that nature is secretly arithmetic, but that the branching rules and the physics of flow impose constraints. Gravity pulls water downhill; slopes and soils decide where channels can persist; erosion and sediment transport decide which incisions survive long enough to become permanent. Over time, many local accidents are edited away. Some channels capture neighbors, some are abandoned, some are buried by landslides or filled by sediment, some are deepened into durable paths. The network that remains is the survivor of countless small competitions for flow and for the power to move sediment. Horton's ratios are a way of summarizing the outcome of that competition.

There is also a practical reason Horton's ordering spread: it makes rivers legible to hydrology. A storm does not merely add water to a basin; it activates pathways. Headwater channels can deliver quick pulses, while larger channels integrate them, smoothing some variation and amplifying others. If you know how many first-order streams there are, how long they tend to be, and how they connect upward in order, you can begin to anticipate how fast a basin might respond to rain, how sharply a flood wave might rise, and how far the influence of a storm might propagate downstream. The hierarchy becomes a rough wiring diagram for moving water.

The historical setting matters here. Horton was writing at a time when field measurements were hard-won and data sets were small. Aerial photography was becoming useful but was not yet the effortless satellite imagery of today. A method that allowed comparison across basins using maps and a few consistent rules was a leap forward. It let researchers talk about drainage networks without relying solely on qualitative description. In that sense, stream ordering was a bridge between the map-maker's eye and the mathematician's notation.

Strahler's later clarifications may seem like a mere technical footnote, but they address a subtle point: a river network can be ordered in more than one plausible way if one is not careful. Horton's original ideas were powerful, yet field application demands consistency. Strahler's formulation made the merging rule explicit in the form most widely used today, especially the "same order makes higher order" rule and the "different orders keep the higher order" rule. That rule has a nice property: it reflects the idea that the main stem's order only increases when it is fed by a tributary comparable to itself, not when it simply absorbs a smaller side stream. In other words, the trunk earns its next rung only when it meets a peer.

There is a quiet philosophical move embedded in that rule. It implies that the significance of a confluence is not determined solely by discharge at that moment. A small tributary can carry a lot of water during a particular storm, yet in the hierarchy it remains subordinate if its network lineage is shorter. The ordering system is not an instantaneous measure of flow; it is a structural description, a memory of how the network is built. That distinction matters because rivers are restless. Flow varies by season and by decade; channels migrate; human diversions alter discharges; climate shifts the timing of snowmelt. The topology of the network changes more slowly. Ordering, when applied to mapped channels, captures a kind of enduring skeleton beneath the changing flesh of water.

A second important idea in Horton's approach is the separation of *scale* from *shape*. If you measure the total length of streams in a basin, big basins will obviously tend to have bigger totals. But if you look at ratios across orders, you get dimensionless numbers that can be compared across basins of different size. Horton emphasized ratios not because he disliked absolute measurements, but because ratios reveal pattern without being drowned by scale. This is also why the ordering system is so naturally paired with other "Horton laws," like the observation that average stream length tends to increase with order by another roughly constant factor, or that average drainage area associated with a given order increases systematically. Taken together, these tendencies suggest that a river basin is not an arbitrary branching; it behaves *as if* it is built by repeated application of similar geometric rules.

That "as if" is where the science becomes honest. No landscape is perfectly self-similar. Bedrock varies, vegetation matters, glaciers carve valleys that do not obey the same rules as rain-fed hillslopes, and humans straighten channels, pave surfaces, and build storm drains that splice new artificial branches into old ones. Even within one basin, mapping choices can shift the count of first-order streams dramatically. Where does a channel truly begin? Is it the first incised gully? The first persistent thread of water? The first place where a blue line appears on a given map scale? Horton's ordering assumes you can identify headwaters, but in the field the boundary between "channel" and "hillslope" is often fuzzy, and it moves with climate and land use.

That fuzziness is not a flaw in the idea; it is part of what makes the hierarchy intellectually useful. It forces a question that hydrology must answer anyway: what pathways are real enough to matter? In a humid forest, a shallow swale may transmit water frequently enough to function as part of the network even if it looks modest. In an arid basin, the same swale might carry water once every few years, and yet during that rare flash flood it may dominate sediment transport.

The hierarchy gives a language for discussing these distinctions, but it does not eliminate judgment. It turns judgment into an explicit choice rather than an unspoken assumption.

A vivid example of Horton's logic appears whenever engineers and planners confront a seemingly small change in the uppermost branches. Consider a suburban development built near the headwaters of a creek. The first-order channels may be buried or rerouted into culverts; small wetlands may be drained; hillslopes may be compacted. To a casual observer the main creek "down below" still exists, so the river seems intact. But from a network perspective, removing or altering a large number of first-order streams is not a minor edit. It changes how quickly water reaches higher orders, how sediment is delivered, and how flood peaks are shaped. In effect, it rewires the capillaries, and the trunk feels that rewiring as a change in timing and intensity. Horton's hierarchy provides a way to articulate why: disturbances at the tips are multiplied by the fact that the tips are numerous and they are the first responders to rainfall.

There is also a human-scale anecdote hidden in many field notebooks: the moment someone tries to "find the source" of a river. Popular imagination wants a single spring or a single point that can be photographed and declared the origin. But in an ordered network, the "source" is plural. A third-order stream does not have one beginning; it has a constellation of first-order origins. Even a modest creek has dozens or hundreds of tiny beginnings that, together, make the larger stream inevitable. Horton's system turns the romantic quest for a single birthplace into a more interesting truth: the river is born everywhere along the divide, in small, repeated acts of gathering.

Once you start seeing rivers as trees, the landscape acquires a different kind of coherence. A drainage divide becomes more than a ridge line: it is the boundary between two competing genealogies. A confluence becomes more than a scenic meeting of waters: it is a merger between lineages whose histories are encoded upstream in the branching pattern. Even the act of naming a stream gains a new layer. Local names tend to attach to prominent segments, but the hierarchy reminds us that prominence is not merely a matter of place; it is a matter of accumulated structure.

Horton's enduring contribution is not that he assigned numbers to streams, but that he made the network countable without stripping it of meaning. The hierarchy preserves the essential asymmetry of river basins: many small contributors feeding a few large collectors. It respects the fact that rivers are directional, that they are built from confluences rather than crossings, and that their organization is not accidental. With that simple arithmetic rule at each junction, a basin becomes an object you can hold in the mind as a whole: a branching logic written into terrain, persistent enough to be measured, and flexible enough to accommodate the roughness of the real world.

3.2 The Fractal Signature

A stream order number is a kind of family rank: it says where a channel sits in the branching tree. Yet two basins can share the same sequence of orders and still feel different on the ground. One may be compact, its tributaries arriving quickly at the main stem; another may be stretched out, with a long central valley and side branches that take their time. Shape matters, not only to aesthetics but to flood peaks, sediment pulses, and the way a landscape remembers storms. The surprise is that shape, too, often comes with regularity.

The clue emerges when mapmakers, hydrologists, and geomorphologists stop staring at individual channels and begin to ask a more abstract question: how does the length of the main river relate to the area that feeds it? That is a relationship any curious person can sense. A creek draining a small neighborhood cannot be a hundred miles long. A continental river draining millions of square miles cannot be a short ditch. But the relationship is not linear, and it is not the simple geometry of circles and squares either. It carries the telltale bend of a power law.

John Hack, working in the mid-twentieth century, looked at real drainage basins and noticed a remarkably consistent scaling pattern: the length of a basin's main channel, call it L, tends to grow with the basin's drainage area, call it A, according to a power law

$$L \propto A^h.$$

The exponent h is now called Hack's exponent, and the relationship is often called Hack's law. It is deceptively compact. The exponent is the whole story: it tells you how quickly rivers "stretch" as the territory they drain expands.

If landscapes were made of Euclidean shapes, the exponent would be easy to guess. Imagine a basin that grows without changing its proportions, like a perfect rectangle scaled up or down. Its length would scale like the square root of its area: double the area and length increases by a factor of $\sqrt{2}$. In that tidy world, h would be $1/2$. Real basins, on average, refuse that simplicity. Hack's empirical work found an exponent closer to about 0.6 in many settings. That number is not sacred, and it is not universal in a strict sense, but it is stubbornly recurring. An exponent larger than $1/2$ implies that basins become, in a statistical sense, more elongated as they grow: the main channel length increases a bit faster than a simple square-root rule would predict.

This is where the phrase *fractal signature* earns its keep. A fractal is often introduced as a shape that looks similar at different magnifications. Coastlines, clouds, and mountain ranges are familiar examples. River networks belong in the same family, but with a caveat: the self-similarity is not exact in the way a mathematical fractal can be exact. Rivers live in dirt and rock, under gravity and climate; they have characteristic scales set by sediment grain sizes, storm footprints, tectonic uplift rates, and the height of hillslopes. Their self-similarity is statistical and approximate, expressed not in identical patterns repeated forever, but in scaling laws like Hack's that hold across a wide range of sizes.

A useful way to picture Hack's law is to imagine zooming out on a map in powers of ten. At the scale of a small forest catchment, the main channel might be a few kilometers. At the scale of a county, it might be tens of kilometers. At the scale of a state, hundreds. If the geometry were Euclidean and proportional, each step up in area would enlarge the basin without changing its "stretchedness." Hack's observation suggests something subtler: larger basins tend to develop a more pronounced spine, a dominant axis that collects flow over longer distances. Tributaries keep branching, but the network's overall footprint often becomes more elongated than a simple scaled-up miniature of a small basin.

There is a human reason this matters. Think about flash floods versus slow-rising floods. A compact basin can deliver rainfall to its outlet quickly, because many points are relatively close in travel time. An elongated basin can spread arrivals out, because the farthest headwaters are much farther from the outlet along the channel network. Basin shape is thus a first-order control on hydrographs, the time series of discharge during and after storms. Hack's law, by relating shape to size, offers a route from a map to an expectation about timing. Even without computing travel times in detail, one can sense that a river draining a large area with a long main channel is not merely "a bigger version" of a small creek; it can behave differently because the geometry organizes delays.

Hack's own period of work also helps explain why the law became influential. Mid-century geomorphology had a hunger for quantitative relationships that could be checked against data. River basins were one of the few landscape features that came with ready-made maps and a clear functional role: they collect and convey water. A law relating length and area offered a handle on basin shape, a way to bring measurement to something previously described with adjectives. It also fit naturally with Horton's earlier impulse: if branching trees have regularities, perhaps basin outlines and channel lengths do as well.

Yet the most instructive feature of Hack's law is not the elegance of $L \propto A^h$, but the fact that the real world both respects it and resists it. Plot L against A on ordinary axes and you see a cloud of points. Plot them on logarithmic axes and the cloud often collapses into a band that looks roughly like a line. The line is the signature of a power law; the thickness of the band is the signature of geology, climate, history, and measurement choices. Some basins are longer for their area, some shorter. Some have more tortuous main channels. Some sit in mountains where slopes force channels to incise quickly and run straight down valleys; others wander across broad plains. Even the definition of "main channel length" is not entirely innocent: do you follow the longest flow path? the path of greatest discharge? the channel that carries the name? Different conventions can shift L by enough to matter.

Those ambiguities have kept the topic lively. Later researchers, including Peter Sheridan Dodds and Daniel Rothman (2000), emphasized that scaling laws in river networks come with substantial fluctuations, and that those fluctuations can grow with basin size. Put differently: larger basins are not only larger; they are more diverse in the ways they can be large. A small catchment has limited options for its overall planform. A continent-scale basin can include mountain belts, plateaus, lowland fans, and glaciated headwaters in one stitched-together story. The exponent you estimate can drift depending on which range of scales you sample and how you define the objects. Hack's law is a powerful organizing principle, but it is not a dictator.

A concrete example helps pin the abstraction to something tangible. Compare two rivers that many people have seen on maps: the Susquehanna and the Potomac in the eastern United States. Their drainage areas are of the same order of magnitude, and both ultimately deliver water to the Atlantic, yet their outlines and main-stem geometries differ. The Susquehanna's basin stretches far into upstate New York and central Pennsylvania, with a long north—south sweep; the Potomac is more constrained by the Appalachians and the Piedmont, with a different orientation and confinement. Even without quoting numbers, one can see that the "spine" of the Susquehanna is unusually long for the footprint it drains. In Hack's language, it is a basin with a relatively large coefficient in the scaling relationship, an elongated form expressed in a longer main channel length relative to area. That elongation shows up in flood timing and in how tributary floods synchronize (or fail to synchronize) at key confluences.

Here a subtle but important point enters: Hack's law contains both an exponent and a prefactor. Written more fully, it is often expressed as

$$L_\ell = C\,A^h,$$

where C is a coefficient that depends on units and on the characteristic "stretch" of basins in a given setting. Modern large-scale analyses have argued that much of the environmental imprint may appear more strongly in C than in h: basins in different tectonic or climatic regimes can be systematically longer or shorter for their area, while the exponent remains comparatively stable across many regions. That distinction is a gift to intuition. The exponent captures something like a general rule of network geometry, while the coefficient carries local personality: rock type, relief, and the inherited scars of glaciation or uplift.

Why should any exponent near 0.6 appear so often? This is the kind of question that tempts people toward mystical universality. A more grounded answer comes from thinking about constraints and compromises.

A river network has competing tasks. It must efficiently collect water and sediment from a two-dimensional surface and funnel them into one-dimensional channels. If channels were perfectly straight and basins perfectly compact, collection would be fast but the network would require many long channels to reach everywhere. If channels were extremely sinuous and basins extremely elongated, a single long spine could reach far, but tributaries would have to travel awkward paths, and the network might waste length or fail to drain efficiently. Real landscapes sit between extremes. Over geological time, channels migrate, capture, and adjust their slopes; hillslopes deliver sediment that can choke channels or force them to incise; storms impose variability in discharge that punishes fragile configurations. The resulting network is a compromise shaped by physics, not a geometric toy. A power law exponent can be the numerical shadow of that compromise.

Another way to phrase the same idea is to ask how a network fills space. A drainage basin is an area, but the drainage lines are a network embedded within it. In a simple Euclidean world, a line does not "fill" an area. In a river basin, the network spreads out to reach almost everywhere, because water finds any weak path downhill and erosion amplifies it. The network is not area-filling in the strict mathematical sense, but it is far more pervasive than a single line. It is this in-between character that fractal language tries to capture: a network that behaves like a tree, but whose branches are numerous enough to imprint the entire landscape.

Yet there is a danger in treating Hack's law as a license to declare river basins "fractal" and be done. Fractals in mathematics can have self-similarity down to arbitrarily small scales. Rivers cannot. At very small scales, you run into grains, roots, soil structure, and the reality that flow may be sheet-like rather than channelized. A rill is not simply a scaled-down river; it can be created and erased within a single season. At very large scales, you run into the curvature of Earth, changes in climate across latitudes, and tectonic architecture that imposes boundaries not present in small basins. Scaling laws often hold in the middle, across the range where channels are persistent and the landscape has had time to organize, but they can bend at the ends.

This is not an inconvenience; it is a clue about what the law is really measuring. Hack's law is less a statement about perfect self-similarity than a statement about *how river basins depart from simple similarity*. The exponent larger than $1/2$ encodes the tendency toward elongation with scale. The scatter encodes the variety of ways landscapes can realize that tendency. Together they suggest that there is a common geometry at work, but it is filtered through local conditions.

The question of local conditions has produced competing explanations. One family of ideas emphasizes topology and statistical self-similarity: if the branching structure follows certain probabilistic rules, perhaps Hack's law follows as a consequence. Another family emphasizes physics in disordered materials: water does not flow across an ideal plane but through a medium riddled with heterogeneity, from fractures in bedrock to patches of clay and sand to vegetation that changes infiltration. Allen Hunt (2016), for example, used ideas from percolation theory to argue that the range of Hack exponents can be understood as a reflection of least-resistance paths through heterogeneous landscapes. In that picture, a river is not simply carving a channel where it pleases; it is discovering and amplifying the easiest connected routes through a complicated substrate. The exponent is then not a universal constant but a statistical consequence of disorder, with variability that is real, not merely measurement error.

Those perspectives need not be enemies. A landscape can be both constrained by branching logic and shaped by heterogeneity. In fact, much of the fascination lies in the way large-scale regularities emerge despite small-scale messiness. A river basin is one of the rare natural systems where you can trace a path from microscopic irregularities to macroscopic order with your eyes. A raindrop hits soil. Microtopography steers it a few centimeters. Those centimeters, repeated over billions of drops and thousands of storms, deepen a rill. Rills connect. Confluences stabilize. The network becomes a memory of where water tends to go.

Hack's law adds a further twist: the memory has a particular scaling. If you take two basins that differ in area by a factor of one hundred, the larger basin's main channel is not ten times longer (as $h = 1/2$ would suggest), but often closer to sixteen times longer if $h \approx 0.6$, since $100^{0.6} \approx 16$. That is a striking departure, large enough to matter for travel times, sediment residence, and ecological connectivity. Fish migrations, nutrient spiraling, the spread of pollutants, and the pacing of flood waves all live in those exponents and coefficients, even if they were first noticed as a mapping curiosity.

There is an intimate aesthetic dimension as well. Many people who hike along rivers feel that a small creek has the "same kind of shape" as a major river, only reduced. That feeling is not entirely illusion. The repeated angles of tributary junctions, the tendency for valleys to branch, the way a main stem collects a constellation of side channels: these are motifs that persist across scale. Hack's law offers a quantitative counterpart to that aesthetic recognition. It says: yes, the motif repeats, but it repeats with a systematic distortion, a gentle stretching that becomes more pronounced as the network grows.

On a map, that stretching can be hard to unsee once it is pointed out. Small basins often look like rounded leaves; large basins often look like long feathers with a shaft and barbs. The feather is not drawn by a designer; it is etched by countless flows, each one following gravity but also constrained by the work already done by earlier flows. The main channel becomes a line of inheritance: once established, it attracts tributaries, captures flow, and deepens its claim. In that sense the basin's elongation is not just geometry, but history. The exponent is a way of writing that history in a single, compact number.

A river network never achieves perfect self-similarity, but it can achieve something more interesting: a kind of disciplined variation. The landscape permits many realizations, yet it nudges them into a family resemblance that can be measured. Hack's law is one of the clearest expressions of that resemblance. It is a reminder that scale is not merely a matter of size; it changes the very proportions of the object, as though the landscape were quietly editing its own blueprint as it grows.

3.3 The Architecture of Efficiency

A river network can be read two ways. One reading treats it as a historical record: storms, landslides, uplift, and vegetation changes leave marks that accumulate into a pattern. The other reading treats it as a solution to a problem. Water must get off the hills. Sediment must be carried away, at least often enough that channels do not choke themselves. Gravity provides the drive, but gravity does not dictate the blueprint. The landscape has choices: many possible branching arrangements can drain the same basin. When a pattern repeats across continents, the mind naturally asks whether those choices are being filtered by a kind of preference, as though the terrain were selecting designs that work better.

The word *efficiency* is slippery. It can mean different things depending on who is speaking. For a hydrologist worried about flood peaks, an "efficient" network might move water to the outlet quickly, producing sharp hydrographs. For an ecologist, it might maintain connectivity while providing refuges and slow zones. For a geomorphologist, it often points toward energy: the work a river must do to overcome friction and to lift and move sediment. Friction turns the potential energy of elevated water into heat; sediment transport turns it into rearranged grains and eroded banks. Either way, energy is spent, and if there is a tendency toward certain geometries, one natural hypothesis is that those geometries spend energy in a particularly economical way.

This idea has a long pedigree in physics and biology. Branching structures are everywhere because they solve transport problems: blood vessels deliver oxygen, lungs deliver air, tree roots deliver water, fungal mycelia explore soils, lightning delivers charge. In the late twentieth century, network scientists and physicists began writing down general "least-cost" principles that can produce branching networks under broad conditions. In one influential vein of work on efficient transportation networks in the 1990s, a simple premise recurs: if many sources must deliver something to a sink (or a sink must collect from many sources), building and maintaining pathways costs something, and moving material along them costs something, then the best design is rarely a single straight line and rarely a uniform grid. Branching becomes the compromise between construction cost and transport cost.

Rivers add a twist that makes the problem more vivid: their pathways are not built in advance. The river *is* the builder, and it builds with the very energy it is trying to dissipate. When water concentrates into a channel, it flows faster, cuts deeper, and becomes more persistent, which concentrates flow further. A rivulet is a self-fulfilling prophecy. This feedback makes rivers seem almost purposeful, though the purpose is simply the physics of erosion responding to concentrated shear stress on the bed and banks.

The hypothesis that river networks behave like energy-minimizing machines found a particularly sharp expression in the idea of *Optimal Channel Networks* (often abbreviated OCNs). Riccardo Rigon, Andrea Rinaldo, and collaborators in the early 1990s gave the approach a clear computational and conceptual form. Instead of trying to simulate every raindrop and every grain of sand, they posed an abstract problem: given a surface that must drain to an outlet, what branching network minimizes a measure of energy dissipation, subject to the constraint that water must be carried downhill along the network? When they let a computer search for networks that score well under the chosen energy measure, the resulting patterns looked uncannily like real drainage networks: dendritic, tree-like, with scaling properties reminiscent of the empirical laws.

At first, the idea can feel too neat. Real landscapes have bedrock ridges, faults, landslides, and vegetation. They are not free to draw any network they please. Yet that objection is also part of the appeal. An optimal-network model is not claiming to reproduce every bend in a specific river. It asks whether, amid all the geological particulars, there exists a general tendency that nudges networks toward a recognizable family of forms, the way evolution nudges many unrelated organisms toward similar solutions (wings, eyes, streamlined bodies) when the physical problem is the same.

To understand what "energy dissipation" means in this setting, imagine a thin slice of water moving downhill. It starts with potential energy because it is elevated. As it flows, it loses that energy to friction with the bed and banks and to turbulence within the water. The steeper the slope and the rougher the channel, the faster energy is dissipated per unit distance. Now add sediment. Moving sediment requires work: grains must be lifted, rolled, or bounced. If the flow is too weak, sediment deposits and can raise the bed, forcing the river to adjust. If the flow is too strong, it can scour and incise. Over long times, rivers tend to arrange their slopes and widths so that they can, on average, move the sediment supplied to them. That tendency is sometimes described as a kind of dynamic balance, and it makes the network's geometry inseparable from the energy budget.

In an OCN framework, one typically compresses this complexity into a cost functional, a mathematical expression that assigns a "price" to a network. The precise form varies, but the spirit is consistent: penalize long paths, penalize steep drops, penalize concentrating too much flow in too many places, or penalize dissipating too much energy overall. Then search for the network that minimizes the price. The network is represented as a tree on a grid or on an abstract surface, and the "flow" is assigned by the contributing area draining to each link. The search can be done by rearranging connections and accepting changes that lower the cost, a process that resembles annealing in physics, where a material is cooled slowly so it can settle into low-energy configurations.

There is an immediate conceptual reward: it explains why branching is so common. Compare two extreme drainage strategies in a basin. In the first, water is forced into a single straight channel early, like a deep ditch running down the middle. Many hillslope points would then have to deliver water laterally to that ditch, possibly across long distances. In the second, the basin is threaded by many channels, each one close to a hillslope point, like a dense comb. That reduces lateral travel distance, but it creates many channels that must be maintained, each one a site of frictional energy loss. Branching networks sit between these extremes: enough channels to reach the hillslopes without excessive lateral travel, but not so many that energy is wasted maintaining a redundant grid.

A concrete analogy makes the compromise more intuitive. Picture an old city with a main railway station and many small neighborhoods. You could give every neighborhood a direct rail line to the station. That would be convenient, but expensive and tangled. Or you could give only a single line running past the city center and force everyone to walk to it. That would be cheap but slow and crowded. Real cities create hierarchies: local streets feed into collector roads, which feed into arterials, which feed into highways. That is not an accident of planning taste; it is the logic of moving many small flows into fewer big flows while controlling cost. River networks are the same logic enacted by water and sediment.

A river basin offers an especially poignant illustration because human infrastructure often collides with its efficiency. Consider a mountain valley where a road runs along the river. The river seems to "choose" the valley floor because it is the easiest route downhill, and the road chooses it for the same reason. But the river's efficiency is built around occasional floods that occupy the whole valley bottom, rearranging gravel bars and eroding banks. A road demands stability, a denial of that occasional rearrangement. Engineers respond by straightening the channel, armoring banks, and confining the river to a narrow corridor. For a few decades, it can look like a triumph of control. Then a rare storm arrives. Water, denied access to its broader floodplain, rises higher, runs faster, and concentrates energy in fewer places. The channel may incise, bridges may scour, and the system can fail in a way that feels sudden but is actually a direct consequence of forcing an energy-dissipating network into a geometry that no longer distributes energy as it evolved to do.

This brings us to an important nuance that physicists highlighted when applying optimization to rivers: the "best" configuration depends on what is being optimized and over what constraints. In some analyses of optimal channel networks, including work associated with Marco Colaiori and colleagues in the 1990s, the global minimum of a chosen energy functional did not always look most like a real river. Networks that were trapped in local minima, configurations that were not the absolute best under the abstract rule but were hard to escape via small changes, often resembled real drainage patterns more closely. That is a deeply natural outcome. Landscapes do not get to reshuffle their entire network at once. They adjust through incremental changes: a meander migrates, a tributary captures another, a landslide blocks a channel and forces rerouting. Such a system is more likely to settle into a "good enough" configuration shaped by history than into a global optimum that requires wholesale rearrangement.

The idea of local minima also helps reconcile efficiency with contingency. A basin can be efficient in the sense that it does not squander energy, while still bearing the marks of accidents and inheritance. A glaciated valley may impose a straight trunk that is not the ideal form for the current climate. A fault line may align tributaries. A resistant bedrock ridge may force a river to take a longer route around it, increasing length relative to area. The network can still exhibit scaling laws and hierarchical structure because those patterns are robust to moderate distortions. Efficiency is not a commandment; it is a statistical tendency that survives imperfect conditions.

The language of optimality can also clarify why Hack's law and stream ordering are so often observed together. In a purely Euclidean scaling world, basin length and area would maintain proportionality, and stream orders would merely count nested mergers without implying anything about basin outline. In an efficiency-filtered world, larger basins can become more elongated because elongation reduces redundancy in long-distance transport. A long trunk can collect flow from far headwaters without requiring an excessively dense network of long channels. Side branches can remain relatively short compared to the trunk. The result is a geometry that naturally supports a power-law relationship between length and area with an exponent that differs from $1/2$, and a branching hierarchy where the ratios of stream numbers and lengths across orders become approximately regular.

If this sounds abstract, it becomes palpable when you look at a landscape where the constraints are simple and the imprint of optimization is easier to see. Volcanic islands offer such examples. On a young volcanic cone with relatively uniform rock and a roughly radial slope, channels begin as small rills and rapidly organize into dendritic patterns around the cone. The network has no inherited fractures or folded strata guiding it strongly. Gravity and erosion do most of the work. Over time, gullies deepen, headward erosion lengthens channels toward the summit, and tributaries join into larger valleys. The outcome is a textbook-like branching structure that seems to "solve" the problem of draining a symmetric surface. In such settings, even casual observers can see the logic: the channels spread out to cover the cone while maintaining a hierarchy, like veins on a leaf.

The same logic appears at much smaller scales, where one can almost watch optimality being negotiated. After a wildfire strips vegetation from a steep slope, a heavy rain can etch a new rill network in hours. Initially, the pattern is messy. Rills form wherever runoff concentrates, often guided by microtopography. Then the feedback begins. A slightly deeper rill captures more flow, erodes faster, deepens further, and steals water from its neighbors. Some rills become the "winners" and persist; others are abandoned and fade. If you return after several storms, the surviving network often looks less random than the first day's pattern, as though the slope had been editing it toward a more stable configuration. It is not solving equations, but the feedback between concentrated flow and erosion performs a kind of search through possible configurations, selecting those that persist under repeated forcing.

There is a temptation to treat optimization claims as either profound truths or empty metaphors. Rivers encourage a more careful middle ground. Energy-minimizing models succeed because they capture a real physical tension: water tends to follow paths that reduce resistance, and once a path is slightly favored it becomes easier still. But they cannot be taken as literal blueprints because they omit much of what makes rivers rivers: unsteady flows, sediment pulses, vegetation, ice, and human interventions. Their value lies in showing that a small set of principles can generate network forms with the right family resemblance, and in giving a language for why that resemblance persists even when details vary.

One of the most practically useful consequences of thinking in terms of energy is the shift from "shape as scenery" to "shape as budget." A bend in a river is not only a curve; it is a place where energy is redistributed between erosion on the outer bank and deposition on the inner bank. A confluence is not only a meeting; it is a mixing of momentum, sediment loads, and bedforms. A braided reach is not only a tangle; it is often a signal that the river is overloaded with sediment relative to its ability to transport it, spreading flow among multiple threads to dissipate energy and to sort grains. The network scale adds another layer: the placement of tributaries and the length of the trunk influence how energy is delivered downstream over time, shaping where channels incise, where they aggrade, and where floods do their most transformative work.

Efficiency, in this sense, does not mean minimizing floods or minimizing erosion. Rivers are not designed to protect roads or to preserve banks. Efficiency means something closer to *compatibility*: a geometry that can, over the long term, accommodate the work imposed by gravity, climate, and sediment supply without constantly reorganizing itself catastrophically. When the work changes—a new dam traps sediment, a city paves hillslopes, a warming climate shifts snowmelt timing—the existing geometry may become less compatible. The river then seeks a new arrangement, sometimes gently through gradual migration, sometimes violently through avulsion or incision. Those episodes are not failures of the river; they are the river rebalancing its energy accounts.

The deeper beauty of the efficiency hypothesis is that it invites a kind of humility. A river network is not merely a set of lines to be controlled, nor merely a passive product of the rocks beneath it. It is a dynamic structure that carries water and sediment with a geometry that has been tested by countless storms. The branching pattern is a compromise written into terrain, and the compromises are legible if one learns how to read them: in the spacing of tributaries, in the elongation of basins, in the way small channels feed larger ones without endlessly proliferating. Stand at a confluence after rain, watch the two colored flows braid together and then settle into a single thread, and it is hard not to feel that the landscape has learned, over time, how to spend gravity's gift without wasting it.

Chapter 4

Liquid Sandpaper

Pure water flows over rock without leaving a scratch; it is the rock fragments carried within the flow that turn a river into an industrial cutting tool. By tracing the journey of a stone from a crumbling hillside to the riverbed, we uncover the physics that allows a fluid to carve canyons through solid granite.

4.1 Arming the Current

A river can be as wide as a football field and as loud as a freight train, and still be strangely gentle to bare rock. Pour a bucket of clean water onto a granite slab and it will shine and run off, leaving the slab essentially unchanged. Even a steady sheet of water, by itself, mostly *slides*. To carve a valley, to bite into bedrock, to plane down a ledge, the current needs something else: a set of hard edges, carried and wielded by the flow. The landscape supplies those edges in the form of sediment—sand, gravel, cobbles, boulders, and the invisible chemistry of dissolved ions. Without that material, water is a poor sculptor. With it, water becomes liquid sandpaper.

That phrase matters because it contains a quiet reversal. In everyday language we imagine a river as the actor and the sediment as the victim: the river *takes* sand and gravel away. In the mechanics of erosion, the sediment is often the active agent. It is the grit that does the scraping. The water is the muscle that lifts, accelerates, and slams grains into rock faces. If the grains disappear, the river may continue to flow, but its ability to cut down and widen can collapse. So the first question of erosion is not only, "How much water is there?" but also, "What is the water armed with?"

The armory is built upstream, on hillslopes and ridgelines, in places that are not rivers at all. Mountains are not worn down by channels alone. They are pre-shattered, pre-weakened, pre-packaged into movable fragments by weathering: the slow, persistent breakdown of rock by frost, heat, roots, acids, salt crystals, and the patient work of microbes. Gravity then takes those fragments and moves them downhill, sometimes in dramatic landslides, sometimes through the nearly invisible creep of soil that inches downslope year after year. Only after this long preparation do particles reach a channel where flowing water can recruit them.

That coupling between hillslopes and channels is one of the most important pieces of river logic: the channel is not an isolated machine that eats rock at a rate determined purely by its own discharge and slope. It is part of a supply chain. A river can run like a conveyor belt, efficiently exporting whatever arrives from the hills, or it can run short of ammunition, polishing its own bed smooth and losing its bite. In the other extreme, it can be overwhelmed with debris, burying bedrock under a protective blanket and again losing its bite. The same material can be a tool and a shield.

Imagine a steep mountain basin after a winter of repeated freeze—thaw cycles. Water seeps into cracks during the day, freezes at night, expands, and wedges the rock apart. A spring storm arrives. On the valley walls, loosened blocks begin to rattle and slide. A debris flow may surge into a tributary, a churning mixture of water, mud, and angular clasts. When that slurry reaches the main channel it does not behave like ordinary flow; it behaves more like wet concrete, thick with stones. When the pulse passes, the channel is left stocked with fresh gravel and cobbles—new cutting teeth. The river has been re-armed in a single afternoon.

On gentler terrain, the supply chain is slower but just as real. In forested hills, soil mantles bedrock like a blanket. Chemical weathering, aided by organic acids, turns minerals into clays, releases ions into groundwater, and rounds grains at the scale of sand. Roots pry open fractures, and burrowing animals churn the upper layers. Over years, the soil gradually thickens and migrates downhill. The movement is not a landslide, not a dramatic event, but a steady, biased random walk. Each rainfall nudges particles a tiny distance. A tree falling on a slope tips up a disk of roots and earth, exposing fresh mineral surfaces and sending a pulse of sediment to a gully. The channel, even in a calm climate, is continuously supplied with fine abrasive grains.

The river uses this supply in different ways depending on the size of the grains and the state of the flow. Coarse gravel and cobbles can behave like a set of hammers and chisels, impacting and plucking at the bed. Sand behaves like sandpaper, especially when grains bounce in short hops (saltation) and strike rock repeatedly. Silts and clays, suspended in the water, are less effective at direct abrasion but are crucial for building floodplains and for carrying nutrients and contaminants. Dissolved ions are the ghostly load: they represent rock that has already been chemically disassembled and is leaving the basin in invisible form. All of these are part of what a river exports. Only some are part of what it uses as weapons.

A striking demonstration of the "tools" idea comes from places where humans unintentionally disarm rivers. Consider sediment-trapping dams. A reservoir is a very efficient settling tank: as water slows, it drops its sand and gravel. Downstream, the released water can be relatively clear—engineers call it "hungry water" because it has spare capacity to transport sediment. Hungry water often erodes banks and beds, scavenging sediment from the channel to make up the deficit. That sounds like an increase in erosive power, and in many cases it is. Yet there is a paradox: if the bed is bedrock, and the dam has intercepted the very gravel that would strike and abrade that bed, the water may be energetic but poorly armed. It can peel away loose material, strip alluvial cover, and even quarry blocks where fractures allow, but its ability to sandpaper intact rock may diminish.

This is not just theory. Downstream of large dams on the Colorado River, researchers and river managers have watched the system shift into a state where sandbars that once refreshed during floods became depleted, while clear releases had the capacity to move sand but not the supply to rebuild it. In the Grand Canyon, debates over controlled floods have often centered on a practical version of "arming the current": if you want flows to rebuild beaches and backwaters, you need sediment available to be mobilized. A flood without sand is mostly a rearrangement of what remains.

Long before modern dams, an older historical episode captured a different side of river weaponry: the extraordinary pulse of debris unleashed by hydraulic gold mining in California in the mid-to-late nineteenth century. Miners used high-pressure water jets—monitors—to blast entire hillsides of ancient gravel deposits, washing the disaggregated sediment into nearby streams. Those streams carried the material into larger rivers like the Yuba and Sacramento. The result was not subtle. Riverbeds rose by meters in places, channels avulsed, and flood risk increased across the Central Valley. Farms and towns found themselves periodically buried in sand and gravel that had not been there a decade earlier. The rivers had been armed far beyond their natural load, but the weaponry did not translate into neat incision; it translated into choking, aggradation, and instability. Sediment was abundant, but the channel could not export it fast enough. Too many tools became cover, and the river's shape was transformed.

That story is useful because it resists a simple moral: more sediment does not automatically mean more erosion. The relationship is closer to a bell curve than a straight line. At very low sediment supply, a river may be starved of abrasive grains. Its bed can become smooth and armored by the few coarse particles it cannot readily move. At intermediate supply, impacts are frequent and the bed is exposed, so abrasion can be rapid. At very high supply, the bed becomes blanketed by alluvium—a layer of sediment that shields bedrock from direct hits. This is the tools-versus-cover tradeoff that mechanistic models of bedrock incision, such as the one Sklar and Dietrich developed in 2004, made explicit: the same grains that do the work can also hide the surface they would otherwise erode.

One way to picture this is to think like a craftsperson sanding wood. A bare plank with no sandpaper stays rough. Add sandpaper and, with steady motion, the surface quickly changes. But bury the plank under a thick layer of sand and you can rub all you like above it; the grains mostly grind against one another. The wood is protected. Rivers face the same logic. A modest gravel supply can be an effective abrasive; a thick gravel layer can act like a mattress.

This tradeoff plays out in bedrock gorges, where the drama of incision is easiest to see. In some canyons, patches of exposed bedrock alternate with wedges of gravel stored behind boulders or in widened reaches. During low flows, gravel sits still and bedrock may be partly exposed but rarely attacked. During floods, the gravel begins to move. Where the moving layer is thin, grains strike bedrock and leave it pocked and fluted. Where the moving layer thickens, bedrock disappears beneath migrating sheets of sediment. An observer standing on a cliff during a flood would see neither process directly; the water is opaque, loud, and fast. The erosion is happening at the interface you cannot see, controlled by how much grit is present and whether it is actually contacting rock.

The source of that grit is often not the main channel itself but the network feeding it: small hollows, landslide scars, and first-order streams that are sometimes dry for months. These headwaters can be thought of as intermittent supply lines. A summer thunderstorm can flush loose material from a gully into the channel. A wildfire can remove vegetation, increase runoff, and trigger debris flows that deliver a sudden cargo of sand and gravel. In the western United States, post-fire debris flows are notorious for arriving with little warning. They are hazardous for communities, but for the geomorphologist they are also moments when a basin is rapidly re-armed. For years afterward, the main channel may work through that stockpile, redistributing it downstream in pulses.

The timing matters as much as the amount. Sediment supply is not steady, and rivers respond to the rhythm. A basin may spend long periods in relative quiescence, with only fine material entering channels, and then experience a single landslide that delivers a century's worth of coarse debris in a day. The river's ability to export that debris depends on its transport capacity—a phrase that sounds like bureaucracy but refers to a simple physical limit: for a given discharge and slope, there is only so much mass of sediment a river can keep moving. If supply exceeds capacity, sediment accumulates. If supply is below capacity, the channel can scour and incise or strip away alluvial cover. This is why a river can flip between being a conveyor belt and being a storage facility.

Because supply is intermittent, storage is not a footnote; it is central to how rivers work. Gravel bars and floodplains are not just scenery. They are warehouses. They hold the very tools the river uses, releasing them during high flows and reclaiming them as waters fall. In gravel-bedded streams, the surface layer can organize into an “armor” of coarser particles, which protects finer material underneath. This armoring can raise the threshold needed to move the bed, changing when and how the river is armed during a flood. A small flow may move sand across the bed but leave the armor intact. A larger flow may break the armor, mobilize coarse particles, and suddenly increase the supply of impacts capable of abrading bedrock downstream.

Human observers have long noticed that rivers “wake up” during floods, and older river lore often attributed this to the water becoming more forceful. That is true, but incomplete. Floods also mobilize stored tools. A river that looks placid in summer may have its bed paved with motionless gravel. When a flood arrives, that gravel becomes a moving abrasive layer. The river is not only stronger; it is armed. The noise of a flood often includes a distinct rumble, sometimes called bedload noise, produced by cobbles colliding and scraping along the bed. That sound is, in a literal sense, the sound of the tools at work.

There is another subtlety: the best tools are not necessarily the largest. A boulder can strike with tremendous force, but boulders move rarely and often only during extreme events. Sand moves much more often, and its cumulative effect can be profound. When sand grains saltate, they hop and collide, delivering repeated impacts. The bedrock surface can develop polish, grooves, and scallops. Over long timescales, many small impacts can outperform a few gigantic ones. This is why desert sandblasting can frost glass; the grains are tiny, but they are relentless. Rivers wield the same principle.

At the same time, sand is easily sequestered. It can be trapped behind large roughness elements, locked into the pores of a gravel bed, or swept into slackwater zones along channel margins. In some regulated rivers, managers measure and track sand budgets precisely because the sand is both the building material for beaches and a key abrasive. A river can carry enormous volumes of water and still be sediment-poor if the upstream landscape is stable and the channels are disconnected from sediment sources. Conversely, a small steep stream can carry a startling sediment load if hillslopes are actively failing into it.

This idea of disconnection—of supply lines being cut—has become increasingly important for understanding how landscapes respond to change. A basin might experience intense weathering on slopes, but if the valley bottom is wide and vegetated, and if tributaries deposit their loads in fans that rarely get reworked, the main channel may receive less coarse sediment than the surrounding terrain produces. The hillslopes are manufacturing tools, but the river cannot access them. Then a climate shift or a land-use change can reconnect the system. Increased storm intensity can incise through fans and terraces, tapping stored gravel. Road building can reroute drainage and trigger gullying. A single avulsion can slice into old deposits. In those moments the river is suddenly re-armed with material that may have been waiting, inert, for thousands of years.

A powerful concrete example sits in many glaciated valleys. Glaciers are remarkable tool-makers: they grind bedrock into a spectrum of debris, from boulders ripped from cliffs to flour-fine silt produced by abrasion under ice. When glaciers retreat, they leave moraines and outwash plains rich in sediment. Rivers flowing from these landscapes can be heavily laden, sometimes milky with suspended glacial flour. The channels are often braided, splitting and rejoining around shifting gravel bars because the sediment supply is high relative to the river's capacity to carry it away in a single thread. Here again, abundant tools do not necessarily produce rapid incision into bedrock; they often produce wide, unstable channels that spend much of their energy moving sediment downstream and sideways rather than cutting down.

These examples point toward an uncomfortable fact for anyone looking for a single "erosion rate" to plug into a model: rivers are not simply erosion machines whose output is controlled by water discharge alone. They are coupled to the physics of weathering and the accidents of slope failure. They inherit the grain sizes and mineral strengths of their basins. They respond to vegetation, fire, and human engineering. The same river can alternate between being a razor and being a blanket, depending on whether it is supplied with the right kind of grit at the right time.

What counts as the "right" kind of grit depends on what the river is trying to do, whether by physics or by human hopes. A river that is cutting a bedrock gorge needs impacts against exposed rock. That favors a supply of sand and gravel that is mobilized but not so abundant that it buries the bed. A river that is building a floodplain and nourishing wetlands needs a steady delivery of fine suspended sediment that can settle in low-velocity zones. A river that is supporting a delta needs both, and it needs them delivered all the way to the coast rather than trapped upstream. When people change the supply chain—by damming, mining, logging, farming, fire suppression, or riverbank stabilization—they are not only changing where sediment goes. They are changing what the river can do.

A final image helps fix the logic. Picture a clear mountain stream flowing over bare bedrock. At low flow it may look like a ribbon of glass, sliding over smooth stone. Now add a handful of sand upstream. The sand disappears, then reappears as a faint hiss along the bottom, each grain bouncing in tiny arcs. The stream has not changed its identity; it is still water moving downhill under gravity. Yet its relationship with rock is altered immediately. The current has acquired edges. It has been armed, not with metal, but with the landscape's own broken pieces, turned into a moving grit that can slowly, inexorably, make stone yield.

4.2 The Threshold of Motion

A river armed with sand and gravel still faces a basic problem: how to get the first grain to budge. A pebble resting on the bed is not waiting politely to be carried away. It is wedged among neighbors, pressed down by its own weight, and cushioned by water that, at low speeds, slips past with little effect. For long stretches of time a stream can look active at the surface while the bed beneath is essentially asleep. Then, with a modest rise in flow—a hard rain, snowmelt in the afternoon, the opening of a dam gate—the bed wakes up. Grains begin to twitch, then hop, then roll in steady motion. The change can feel abrupt, like a switch being thrown. In reality it is the crossing of a threshold: a point at which the forces applied by moving water finally outweigh the forces pinning a grain in place.

The threshold matters because so much follows from it. Below it, a channel tends to store sediment, build armor, and protect its own bed. Above it, the channel can become a conveyor, sending its tools downstream and, in the process, grinding and shaping the bed and banks. If the first section was about how a river acquires tools, this is about whether the river can *use* them.

The physics begins at the scale of a single grain, which is both small enough to hold in your palm and large enough to make a landscape. Picture a rounded cobble nestled against two others. Water moves over it, fastest a short distance above the bed and slower in the thin layer right at the surface where friction drags the flow down. That speed gradient is why the river can push on a grain at all. A moving fluid exerts a *shear stress* on the boundary it slides past: a tangential drag that, at the bed, is transmitted through turbulence and viscous friction to the grains. The word sounds technical, but the idea is familiar to anyone who has rubbed a wet countertop with a sponge. The harder you press and the faster you move, the larger the shear stress.

On a riverbed, shear stress is not applied by a hand but by gravity acting on the mass of water. A river on a slope is perpetually trying to slide downhill. The water's weight provides the pressure; the water's motion provides the drag. When hydrologists and geomorphologists talk about a river having "more energy," they often mean that its shear stress has increased: deeper water, steeper slope, faster flow, or some combination.

The grain resists motion with its submerged weight and with friction and contact forces at its points of support. "Submerged" matters because water provides buoyancy. A stone in air feels heavy; in water it feels lighter because the fluid pushes upward. The resisting force is therefore not the full weight of the grain but the *effective* weight after buoyancy. Add to that the fact that grains interlock. A grain sitting in a pocket is protected. A grain perched on a bump is exposed. Two beds made of the same sediment can behave very differently depending on how the grains are packed.

Water fights back with two principal forces. *Drag* acts roughly in the direction of flow, pushing the grain downstream. *Lift* acts upward, produced by pressure differences around the grain and by turbulent fluctuations in the flow. Lift is easy to underestimate because we imagine the grain being shoved along the bed like a book across a table. But lift is often what breaks contact and reduces friction, turning sliding into hopping. Many grains begin to move not by being bulldozed out of the way, but by being momentarily unweighted—just enough for a drag impulse to pop them into motion.

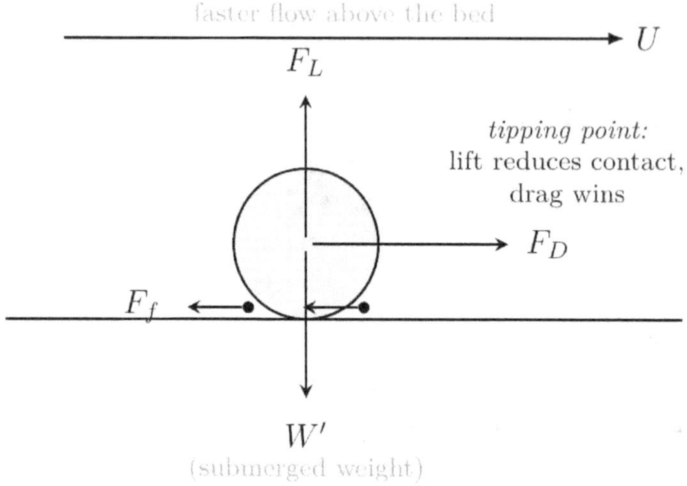

The diagram looks almost too simple for something that can govern the pace of mountain erosion, but that simplicity is the point. At the threshold, the river is negotiating a force balance. If the drag and lift supplied by the flow are, on average, too small, the grain stays put. If they are big enough often enough, the grain begins to move. The tricky part is that "often enough" is doing a lot of work. River flow is turbulent. Even in a seemingly steady current, velocity near the bed pulses and swirls. A grain does not feel one clean, constant push. It feels a jittery sequence of kicks. The threshold, therefore, is not a single moment so much as a change in the odds: below it, the kicks are rarely sufficient; above it, they become frequent.

Engineers and geomorphologists learned early that working directly with all these forces is awkward. Drag depends on the grain's shape and exposure, lift depends on pressure fluctuations, friction depends on packing. A more elegant approach is to compare the *driving* stress applied by the flow to the *resisting* stress implied by the grain's submerged weight. That comparison is made dimensionless, producing the quantity that became famous through the work of Albert Shields in 1936.

The Shields parameter is usually written as

$$\theta = \frac{\tau}{(\rho_s - \rho)gD},$$

where τ is the bed shear stress, ρ_s is the sediment density, ρ is the water density, g is gravity, and D is a representative grain diameter. The denominator is essentially the grain's submerged weight per unit area. When θ exceeds a critical value θ_c, motion begins. The usefulness of this formula is not that it gives a magical universal constant—it does not—but that it tells you what to pay attention to. Increase τ by making the flow deeper or faster, and θ rises. Increase D by using coarser sediment, and θ falls. Put differently: the threshold is easier to reach for small grains than for big ones, unless the small grains are sheltered.

If you have ever watched a child at a beach try to get a stone moving in the swash zone, you have seen a crude version of Shields' logic. A thin sheet of water can move sand easily; it struggles with pebbles. Increase the wave energy and suddenly pebbles tumble too. Rivers do the same, except their energy is set by slope and discharge rather than by incoming waves.

The Shields framework also hides a second control: the nature of the flow right at the grain scale. Water can behave in a more streamlined way around small grains, or in a fully turbulent way around large roughness elements. The transition is captured by a particle Reynolds number, a measure of whether viscosity or inertia dominates in the flow around the grain. This is why thresholds are not identical for silt, sand, and cobbles even after you non-dimensionalize. A smooth grain in a gentle, viscous-dominated flow can be surprisingly resistant. A rough grain in fully turbulent flow may be easier to mobilize than you would guess from weight alone.

All of this sounds tidy until you meet a real riverbed. Natural beds are mixed. A gravel bar is not a laboratory tray filled with uniform spheres. It is a messy mosaic of sizes and shapes, including flat stones that behave like roof tiles, angular fragments that lock together, and pockets of sand that hide between larger clasts. The same shear stress that would easily move a loose sand bed might not move sand that is tucked into the pores of a gravel framework. In that case the sand is not free to be picked up; it is sheltered. Geomorphologists call this *hiding*. Conversely, the larger grains can be more exposed to the flow, sticking up into faster water. That exposure can partially compensate for their weight. The result is that the critical condition for motion becomes grain-size dependent in a complicated way even under the same flow.

This mixing produces one of the most consequential behaviors in gravel-bed rivers: armoring. During moderate flows, finer grains tend to be transported more readily. Over time, the surface can become enriched in coarser particles, which form a protective pavement. The riverbed then requires a larger shear stress to mobilize than you would predict from the average size in the subsurface. A casual observer might say the river has "run out of sediment" because the bed looks stable. What has really happened is that the bed has organized itself into a state that resists motion.

At this point it becomes clear why compilations like Buffington and Montgomery's 1997 review of incipient motion studies found not one threshold but a family of them. Different experiments define "motion" differently. Is the threshold the first visible twitch of a grain? The first grain that travels a diameter downstream? A small but measurable transport rate? Each definition gives a different number. The bed's history matters, too. A freshly disturbed bed is loose;

a bed that has had time to settle and pack is tight. In rivers, floods can reset the packing, then low flows can allow it to stiffen again. The threshold is not simply a property of the water and the grain; it is also a property of the bed's memory.

There is a valuable historical anecdote hidden inside this measurement problem. In the early twentieth century, river engineers wanted reliable formulas for sediment transport because harbors silted up, canals filled, bridges undermined, and reservoirs lost capacity. They built flumes, ran water over sand, and watched when grains started to move. The experiments were painstaking, but the results did not collapse neatly onto a single curve. Shields' 1936 work was influential partly because it offered a unifying dimensionless approach at a time when practitioners were drowning in site-specific numbers. Yet even then, the curve was not a law like the boiling point of water. It was a practical synthesis of messy reality, useful precisely because it summarized variability rather than pretending it did not exist.

Another way to make peace with the mess is to abandon the idea of a sharp switch and treat the threshold as probabilistic. Hans Albert Einstein, working on bed-load theory around 1950, leaned into this view by treating transport as the net outcome of grains randomly entering and leaving motion. Even if the average shear stress is below a nominal threshold, a turbulent burst can move a grain for a short distance. Even if the average shear stress is above it, a grain might be trapped in a pocket and remain still. In this picture, increasing flow does not flip transport from zero to one. It increases the fraction of time grains spend moving, increases the length of their hops, and increases the number of grains recruited into motion.

This probabilistic perspective matches what you can see with your own eyes if you linger by a stream during rising flow. The bed does not start moving everywhere at once. First there is sporadic clicking as an occasional pebble shifts. Then a few grains begin to travel, stopping and starting. Only later does the motion become continuous enough that you stop noticing individual clasts and start noticing the collective behavior: bars reshaping, patches of bed scoured clean, a faint cloud of sand lifting into the flow. The transition from stability to transport is sharp enough to be dramatic, but fuzzy enough to be interesting.

The threshold also introduces a powerful nonlinearity into river response. A small increase in discharge can produce a disproportionate increase in sediment transport once the threshold is crossed. This is why floods are so geomorphically important. The everyday flows that maintain aquatic habitat and carry dissolved ions may do relatively little rearrangement of coarse sediment. A rare event that lasts a day or two can move more gravel than several years of ordinary flow. That unevenness is not just because the flood has more water. It is because the flood crosses thresholds that the daily flow never reaches.

Engineers sometimes summarize the idea with a deceptively simple rule of thumb: "Most sediment moves on a few days each year." The truth varies with climate and geology, but the logic is robust. Near the threshold, transport is intermittent and sensitive. Far above it, transport can become limited not by whether grains can move, but by how much sediment is available and how the bed organizes itself. A channel that is slightly below threshold is a storage zone; a channel slightly above threshold can become a highway.

A concrete, everyday example comes from urban streams after a storm. A small creek running through a park may appear harmless on a dry day, barely able to push aside leaves. After a heavy downpour, storm drains route runoff quickly into the channel. Water level rises, velocity increases, and the shear stress on the bed can jump. Gravel begins to move, sometimes audibly. If the creek has been channelized, straightened, or confined, the increased slope and reduced roughness can further increase shear stress, pushing the bed over threshold more often. The result can be sudden bank erosion and bed scour that surprises residents: the creek looks the same size, but it has crossed into a different mechanical regime. The threshold has been crossed not by a change in the grains, but by a change in how the city delivers water.

There is also a deeply human resonance in the idea of threshold because it separates two kinds of risk. Below threshold, the bed is stable; bridges and pipelines rest on a relatively fixed foundation. Above threshold, the foundation becomes mobile. Scour around bridge piers is, in essence, a local crossing of the threshold caused by flow acceleration and turbulence around the structure. Many bridge failures during floods have been caused not by the water pushing the bridge sideways, but by the riverbed being excavated from beneath it. The bed did not merely get wet; it started behaving like a moving granular material.

If the threshold is the gate that allows sediment to move, it is also the gate that allows sediment to *work*. A grain that never leaves the bed cannot strike bedrock, cannot abrade, cannot help carve a pool. The same shear stress that initiates motion is also a prerequisite for the river to deploy its sandpaper. This creates a striking sensitivity: a river can be fully armed with sediment, and still do little erosive work for long periods if it rarely crosses the threshold. Conversely, a river with modest sediment supply can do intense work during rare events if those events reliably mobilize the available tools and keep the bed exposed.

The most vivid mental image of the threshold is not a number or an equation, but a single grain at the moment it hesitates between rest and motion. Water pushes; turbulence flickers; the grain rocks in its pocket. For a second it lifts, barely, and in that tiny loss of contact friction collapses. What was wedged becomes free. The grain rolls half a turn, then hops, collides with another grain, and the collision propagates. One moving clast strikes another and, in a rising flood, the bed's stillness can unravel into widespread motion. The river has not changed its purpose. Gravity still pulls water downhill. But a boundary has been crossed, and the bed has shifted from solid ground to a granular swarm that can rasp, batter, and reshape whatever lies beneath.

4.3 Ghosts, Clouds, and Rolling Stones

Stand beside a river and you can often read its sediment like a mood. After weeks of dry weather, the water runs clear enough to show every pebble on the bed. After a storm, it turns the color of coffee-with-milk, opaque and impatient. In spring snowmelt it may carry a steady hiss of sand along the bottom while the surface looks almost serene. These are not cosmetic changes. They are the river expressing three different ways of carrying rock away: as invisible chemistry, as a cloud in the water column, and as stones that roll and bounce along the bed. Each mode has its own physics, its own tempo, and its own kind of landscape-making.

The first mode is the hardest to see, and in some sense the hardest to feel, because it is *absence*. Rock can leave a mountain without ever becoming a grain. Rainwater absorbs carbon dioxide from the air and soil, forming a weak acid. It seeps through fractures, reacts with minerals, and carries away ions: calcium, bicarbonate, sodium, silica. This is the *dissolved load*. It does not make the river muddy. It does not pile into bars. Yet over geologic time it can remove astonishing amounts of material, lowering entire landscapes grain by grain in molecular form. A limestone region, for example, can be eaten from within by dissolution, producing caves, sinkholes, and springs that feel like magic until you realize they are plumbing designed by chemistry.

A classic historical example of this invisible work is the discovery of the "hard water" problem in industrializing cities. Boilers and pipes clogged with scale because rivers draining limestone terrain carried dissolved calcium and bicarbonate; when heated, the dissolved load precipitated back into solid carbonate. The river had not been transporting obvious sediment at all, yet it was carrying the ingredients for rock. The same chemistry that grows stalactites can sabotage machinery. In the landscape, it quietly shifts mass from hills to oceans, where dissolved ions eventually help build shells and marine carbonates. A river can be a conveyor of minerals without ever looking dirty.

If dissolved load is the ghost, suspended load is the cloud. This is what most people picture when they imagine a sediment-laden river: fine silt and clay held up by turbulence, turning water opaque. Suspended sediment gives rivers their color signatures. Glacial streams can be pale and milky from rock flour. Rivers draining volcanic ash deposits can look gray. A flood in an agricultural basin can run chocolate-brown. The suspension is not just dirt in the water; it is the raw material of floodplains, deltas, and many of the soils humans depend on.

Suspended grains stay aloft because they are small enough that their fall speed is slow compared to the upward stirring of turbulence. Gravity is still pulling them down, but the eddies keep catching and lifting them. The river becomes a moving atmosphere in miniature, filled with particles drifting, rising, and settling. When the flow spreads out over a floodplain and slows, the turbulence weakens and gravity wins. Mud settles in thin sheets that accumulate over centuries into rich alluvial soils. The paradox of suspended load is that it is most visible when the river is energetic, but it builds land when the river calms down.

The third mode is the one you can sometimes hear: the *bedload.*

These are the heavier grains—sand, gravel, cobbles—that spend most of their time near the bed, rolling, sliding, or hopping in short arcs. When bedload moves, it interacts directly with the channel's boundary. It is the mode that most strongly controls the geometry of the channel itself: how deep it is, how wide it becomes, where pools and riffles form, how bars migrate. Bedload is also the primary agent of abrasion in many bedrock rivers. If suspended mud is a painter that coats floodplains, bedload is the mason's rasp that shapes the channel.

One reason bedload feels so central is that it is scarce in a particular way. By mass, many rivers export more material in dissolved and suspended forms than as gravel rolling on the bottom. Yet the gravel has leverage. A small flux of moving stones can rearrange the bed and banks far more dramatically than a much larger flux of dissolved ions. In a gravel-bed river, the bedload acts like a skeleton: it defines the channel's roughness, controls how flow accelerates and decelerates, and determines whether the river braids into multiple threads or holds itself to one.

To understand why a river divides into these modes, it helps to focus on a single grain and ask a practical question: if the grain is picked up, will it fall back down quickly, or will it stay in the flow? The answer depends on two competing speeds. One is the grain's settling velocity, w_s, how fast it sinks in still water. The other is a measure of the turbulence and shear near the bed, often represented by the shear velocity u_*, which is tied to the bed shear stress through $\tau = \rho u_*^2$. Turbulence acts like an elevator; settling acts like gravity pulling the elevator cable down.

A compact way to combine these is the *Rouse number*,

$$P = \frac{w_s}{\kappa u_*},$$

where κ is the von Kármán constant (about 0.4). High P means settling dominates: grains spend their time near the bed, moving as bedload or in a near-bed layer. Intermediate P means grains are frequently lifted and can remain in suspension for long distances. Low P means grains are so fine relative to the turbulence that they behave almost like dye in the water—the classic *wash load* that can be transported from far upstream without much interaction with the local bed.

The beauty of P is that it turns an intuitive observation into a predictive one. Increase u_* during a flood and grains that were previously confined to the bed can be lofted into suspension. Decrease u_* during falling flow and the same grains rain out. Change the grain size and the settling speed changes dramatically; a sand grain falls far faster than a clay particle. In one reach of river, the same grain might be bedload on a calm day and suspended load during a flood. Transport mode is not a label attached permanently to a grain; it is a relationship between grain and flow.

This relationship also explains why some rivers have that distinct two-tone behavior in storms. The first flush of runoff often carries fine material—dust, soil, ash—that was sitting on the surface of the watershed. This material is easily suspended, producing a sudden muddy plume. Coarser bedload may arrive later, because it is supplied through different pathways and requires higher shear stress to move. The river's turbidity spikes early; the rumble of gravel may come later, if it comes at all. To a city watching a rising river level, the water turning brown can be an early warning of mobilized soils upstream, even before the bed itself begins to reorganize.

There is a human story embedded in these clouds. In the 1930s, during the Dust Bowl, wind-stripped topsoil was famous for darkening skies and filling houses with grit. Less often remembered is where much of that soil ended up. Rainfall and runoff carried enormous amounts of fine sediment into streams, increasing suspended loads and altering channels. Downstream, reservoirs silted rapidly. Engineers learned that controlling sediment was not only about what a river eroded from its banks; it was also about what a watershed delivered from fields. Suspended load ties river morphology to land use with a tight knot.

Bedload, by contrast, often ties rivers to the drama of thresholds and pulses. Near the threshold of motion, bedload transport is not steady. A grain may sit for hours, then move in a burst of hops, then disappear under other grains, then re-emerge later. Experiments that tag pebbles with paint or electronic tracers have revealed behavior that can feel almost ghostly. A pebble placed carefully on a bar might vanish after a flood, leading you to assume it has been swept far downstream. Months later it may reappear a short distance away, exhumed when the bed scours again. The grain's history includes long rests and sudden flights.

This stop-and-go behavior matters because it makes transport *memoryful*. If sediment moved like smoke diffusing in air, you could predict its spread with a simple bell curve. But bedload often spreads in a way that is too lopsided, with some grains traveling far and others barely moving. Martin and Jerolmack (2012) highlighted how such transport can show "anomalous diffusion": long-tailed distributions of travel distances and rest times. In plain language, bedload does not behave like a polite crowd shuffling forward; it behaves like a collection of individuals with wildly different luck. One grain gets trapped; another catches a fast lane.

That is the "ghosts" part in a second sense: not only dissolved ions, but also grains that disappear into the bed. Riverbeds are porous. Fine sand can infiltrate a gravel framework, hiding from the flow until the framework is disturbed. Even cobbles can be buried by migrating bars. The bed is not a flat conveyor belt; it is a three-dimensional storage medium, constantly sorting and resorting. Burial is not an exception. It is a normal state.

The “rolling stones” part is also literal. On some floods you can hear rocks moving, especially at night when traffic noise quiets. It is a low, continuous rumble, punctuated by clacks when cobbles collide. That sound is mechanical energy being transferred from water to sediment. The same collisions that make noise also chip and round grains. Over long distances, angular fragments become rounded pebbles, not because the river has a preference for circles, but because sharp corners concentrate stress and break off. The river is a tumbling machine.

Now for a twist: even the “cloud” can sometimes act like sandpaper. It is tempting to assume that once grains are suspended, they no longer strike the bed with enough force to matter. Yet work by Lamb and Dietrich (2008) argued that suspended sediment can still contribute to bedrock abrasion if there is sufficient near-bed concentration and if turbulent motions and settling bring grains into contact with the bed. In a strongly turbulent flow, grains do not simply float serenely. They are constantly accelerated by eddies and pulled downward by gravity. They can impact the bed at nontrivial speeds, especially when the bed is exposed and the flow is energetic.

This is important because it changes a common intuition about very large floods. One might imagine that when flow becomes extremely strong, everything is swept into suspension and the bed becomes, paradoxically, protected from impacts. In reality, the transition from bedload-dominated impacts to suspended-load impacts is not a clean handoff. There can be regimes where impacts continue, but with different statistics: more grains aloft, more frequent near-bed contacts, different impact angles. The details remain an active area of research, but the core point is simple: a river can keep abrading even when its tools are not all scraping along the bottom in continuous contact.

The three modes also leave different signatures in the landscape, signatures you can learn to read. Dissolved load leaves behind karst topography, mineral springs, and sometimes the slow loss of relief without obvious debris. Suspended load builds floodplains with layered silts and clays; it produces natural levees and backswamps, and it can bury forests and farms under a skin of mud. Bedload produces bar forms, riffle-pool sequences, step-pool channels in steep streams, and the very texture of the bed that controls habitat for fish and invertebrates.

Consider the Mississippi River, one of the most studied sediment systems on Earth, partly because so many people live on and near its floodplain. The fertile soils of the lower Mississippi valley owe much to suspended load deposited during floods over thousands of years. Yet the shape of the channel, the migration of bends, and the maintenance (or loss) of sandbars involve bedload and near-bed sand transport. The dissolved load, meanwhile, is part of the chemical story of the basin, including nutrient transport that influences the Gulf of Mexico. The river is not one conveyor; it is three overlapping conveyors, each with different consequences.

Or take a mountain stream in the Rockies. Its water can be clear most of the year, with modest dissolved load and relatively low suspended sediment except during storms. Yet a single spring runoff can move enough gravel to rebuild bars, shift the thalweg (the deepest path), and undercut banks. For an angler, that rearrangement changes pools and riffles. For a bridge engineer, it changes where scour might threaten foundations. For the stream itself, it changes roughness, which feeds back into flow patterns and future transport. In small basins, bedload can be the dominant sculptor even if it is the smallest fraction by mass.

There is also a darker, debated mechanism that sometimes enters popular stories about extreme erosion: cavitation. Cavitation occurs when local pressure in fast-moving water drops enough to form vapor bubbles that then collapse violently, producing tiny jets and shock waves capable of pitting surfaces. In engineering, cavitation is notorious for damaging ship propellers and spillways. It is physically capable of rapid erosion under the right conditions. In natural rivers, however, geomorphic evidence suggests it is often minor compared to sediment abrasion. Rivers rarely sustain the extreme velocities and pressure drops typical of engineered nozzles, and the distinctive signatures of cavitation are not commonly documented in the field. It is a useful reminder that water can, under special circumstances, attack rock directly—but it is usually the tools in the water, not the water alone, that do the heavy work.

Seen this way, the river becomes a sorting engine. It takes a mixture of products from weathering and hillslope delivery and divides them into invisible chemistry, mobile clouds, and rolling stones. The sorting is not merely a passive response; it actively organizes the river network. Fine suspended sediment can travel long distances, linking headwaters to lowland floodplains and deltas. Coarse bedload tends to be more local, shaping channels reach by reach, stored and re-released in pulses. Dissolved load integrates the quiet, pervasive chemical conversation between water and rock across the whole basin.

A person watching the water surface might miss most of this. The important action often happens at the boundary: in the thin near-bed layer where turbulence is born, where grains decide whether to rest or move, where a suspended particle either settles or stays aloft, where a buried pebble is either entombed for years or excavated in a single hour. That boundary is where a river turns the broken fragments of a landscape into motion, and motion into form.

Late in the day, when the light is low, a clear stream can look almost empty. Yet it may still be carrying ghosts: a quiet flux of ions, the chemical residue of rock turned to solution. After rain, the same stream may become a cloud, thick with suspended mud that will later settle into a thin skin on a floodplain. In the first hard hours of the rising flow, the bed may begin to speak in clicks and rumbles as rolling stones wake and start their intermittent work. Different modes, one river, each leaving a different kind of mark, each a different way that solidity becomes travel.

Chapter 5

The Winding Path

Water abhors a straight line. This chapter investigates the hidden mechanical forces-from corkscrew currents to sediment overload-that compel rivers to carve intricate, shifting geometries across the landscape, revealing that what appears to be a random wander is actually a precise, self-organizing search for equilibrium.

5.1 The Geometry of the Snake

A meandering river looks, from above, like hesitation made visible: a line that cannot quite commit to straightness. Yet the curves are not decoration. They are the river's working method, a way of spending energy and moving sediment while staying, for long stretches of time, in a recognizable state. The river's apparent whim hides a strict internal choreography, and the lead dancer is a motion most of us never imagine inside flowing water: a slow corkscrew that twists downstream.

The key idea is *secondary circulation.* The main current of a river runs downstream; that is the obvious part. Secondary circulation is the less obvious part: a rolling motion across the channel, as if the water were climbing one bank at the surface and sneaking back along the bed. In a bend, this cross-stream roll becomes organized and persistent, producing a helical (helicoidal) flow that couples hydraulics to geology: it pushes erosion onto the outer bank and lays down a matching deposit on the inner bank. Rivers, in other words, *build the very curvature that steers them.*

Imagine standing on a footbridge above a bend after a rain, watching the surface. The fastest water does not run down the centerline of the valley. It crowds toward the outside of the curve. If you toss in a stick, it accelerates, leans outward, then continues downstream as if pulled along an invisible rail. That rail is not painted onto the water; it is produced by the bend itself. The channel forces the flow to turn, and turning sets up a sideways pressure gradient: water piles slightly higher along the outer bank than the inner, like a thin, moving tilt-table. Gravity tries to level that tilt, pushing surface water inward, while the inertia of the turning flow keeps trying to fling it outward. The argument between these tendencies is resolved by a spiral: near the surface, water drifts outward toward the outer bank; near the bed, it returns inward toward the point on the inside of the bend.

This roll matters because rivers are not just pipes for water. They are conveyors for sediment. Sand and gravel grains along the bed respond primarily to the near-bed flow, where friction is strongest and where the "return" limb of the corkscrew operates. The bottom water, sweeping from the outer bank toward the inner bank, tends to carry bed material inward. Meanwhile, the surface flow, pressing outward, helps position the fastest thread of current near the outer bank. The result is a beautiful, ruthless division of labor: the outer bank is attacked and undercut; the inner bank is quietly padded with new sediment.

That undercut outer bank has a name that sounds like a tool, because in a sense it is one: the *cut bank*. It is commonly steep, sometimes nearly vertical, because erosion there is not a gentle sanding but a bite-and-collapse process. The current scours the toe of the bank, removing the support that holds up the overlying soil. Above, roots and clay may cling for a while, then fail in slabs. Anyone who has walked along a meander bend in late summer knows the feeling: the ground close to the edge is suspiciously crisp, as if it might snap. Across the channel, the inner bank is a different world: a shallow slope of fresh sand and silt called a *point bar*. It looks, and often feels underfoot, like a beach in a place where beaches seem out of context.

The river's bend therefore contains, within a single glance, a small moral drama about time. The cut bank is loss: land eaten away, trees leaning out over air, the past collapsing into the stream. The point bar is gain: new land emerging grain by grain, a platform for willows and grasses to colonize. The planform curve you see on a map is the footprint of this asymmetry. And because the asymmetry is systematic, the footprint tends to persist and even exaggerate.

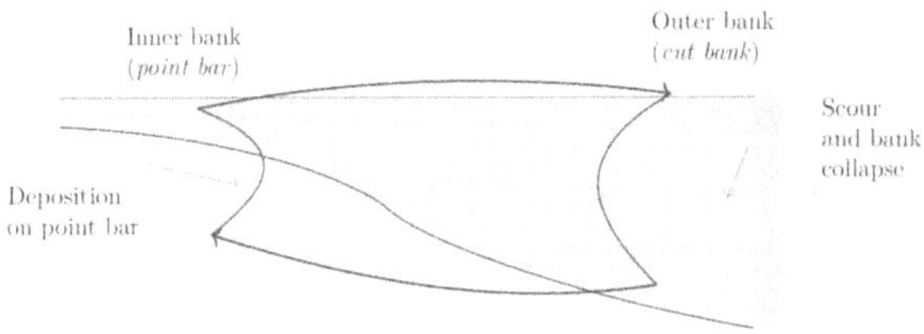

Helical (secondary) circulation in a river bend: surface flow to the outer bank, near-bed return flow to the inner bank.

Once you accept the corkscrew, a meander stops being a mystery and becomes a machine with moving parts. The outer bank erodes, the inner bank grows, and the whole bend migrates sideways and downstream, like a slow handwriting flourish sliding across the page. Migration can be imperceptible day to day, but it adds up. Farmers on floodplains have long known that field boundaries drawn in legal language are not always respected by water in motion. In many regions the law itself had to learn the difference between sudden and gradual change: a boundary that shifts by slow bank erosion and deposition is one kind of problem; a boundary that jumps overnight when the river cuts through a narrow neck is another.

That sudden jump is a *cutoff*. The river's loops, exaggerated by their own internal mechanics, eventually create a geometry too tight to sustain. During a flood, or sometimes just during persistent high flow, water discovers a shorter path across the neck of land separating two adjacent limbs of the same bend. Once a small channel forms, it steals more and more discharge because it is steeper and shorter, and therefore offers less resistance. The old loop is abandoned and becomes an oxbow lake: a crescent of still water that slowly fills with fine sediment and organic matter, turning over years into a marsh, then a damp swale, then a line of darker soil that betrays the river's former position.

The lower Mississippi River provides a famous historical stage for this drama, not because the physics there are exotic, but because humans have been watching and writing about it for a long time. Early travelers described banks "caving" into the current and channels shifting so that familiar landmarks moved relative to the river. By the nineteenth century, river pilots learned that a map was a snapshot, not a promise. The river's bends could migrate in a pilot's working lifetime; cutoffs could rewrite navigation routes on the scale of a single season. The U.S. Army Corps of Engineers eventually undertook large-scale cutoff projects to shorten and stabilize the navigation channel. Those engineering cutoffs were, in a sense, an attempt to do deliberately what the river does opportunistically: trade a long, curving path for a short, direct one. The river's own cutoffs, however, are not motivated by efficiency in the human sense. They are simply the consequence of the river repeatedly intensifying its bends until the geometry offers a tempting shortcut.

The meander has another signature you can see with your feet. Walk from the cut bank to the point bar and you often cross a subtle topography: ridges and swales that run roughly parallel to the inner bank. These are *scroll bars*, former positions of the point bar surface left behind as the channel migrates. Each ridge is a faint shoreline of a past year's deposition, and in some places you can count them like tree rings, though the spacing is irregular because floods do not keep tidy calendars. The soil and vegetation respond. Slightly higher ridges drain better and may host different plant communities than the wetter swales. A meander is not only a curve in water; it is a way of writing a textured ecological template onto the floodplain.

Why do the bends tend to come in a recognizable scale? Not every river looks like a cartoon snake, but many develop a fairly regular spacing of bends, as if some hidden hand were choosing a wavelength. Observations assembled by Luna Leopold and Gordon Wolman in the mid-twentieth century suggested a striking rule of thumb: the distance from one bend crest to the next (the meander wavelength) often falls on the order of ten channel widths, give or take. The number is not magical, and it varies with river type, bank material, and flow regime, but it hints at something important: meandering is not merely a sequence of unique accidents. It has a preferred geometry because the coupled system of flow and sediment has preferred modes of deformation, much as a vibrating string prefers certain notes.

The mechanics behind that preference can be pictured without equations. Start with a nearly straight channel that contains small, random irregularities: a tree root on one bank, a patch of coarser gravel on the bed, a tiny asymmetry in flow where a tributary enters. Any of these can nudge the fastest thread of current slightly toward one side. Once that thread is displaced, it begins to erode that bank a little more, making the channel slightly wider or deeper there. The altered shape then steers the flow even more strongly toward the same side, reinforcing the initial displacement. This is the essence of an *instability*: a small perturbation that feeds on itself. But it cannot grow without limit, because other constraints intervene. Very tight curvature increases friction and promotes cutoff; bank strength can resist erosion; sediment supply can choke a channel or armor its bed. The meander emerges from this contest as a pattern that can persist, migrate, and occasionally reset.

Secondary circulation is central because it provides a reliable way for curvature to reorganize the bed and banks. In a bend, the near-bed flow directed toward the inner bank encourages deposition there, which builds the point bar and tends to push the thalweg (the line of deepest flow) toward the outer bank. The thalweg's outward shift concentrates shear stress on the cut bank, intensifying erosion. In other words, the corkscrew not only moves sediment; it *positions* the strongest flow against the bank most vulnerable to attack. The river sharpens its own tool.

The details can be wonderfully counterintuitive. A casual observer might assume that the outside of a bend is eroded simply because the water "hits" it. That is not quite right. Water does not collide with a bank like a car into a wall; it slides along it, and the most effective erosion comes from sustained shear at the boundary and from turbulent bursts that pluck grains from the toe. The outward shift of high velocity is a consequence of the whole three-dimensional flow field. Likewise, the inner-bank deposition is not merely a matter of the current being slower there, though it often is. It is also a matter of the near-bed return flow delivering sediment inward and of the point bar itself shaping the distribution of depth and velocity, creating a sheltered zone where grains can settle.

Even the word "settle" can mislead. In many rivers, sand on a point bar does not drift down like dust in a quiet room; it is pushed and rolled along the bed in bursts, pausing where local shear stress drops below the threshold needed to keep it moving. Gravel behaves even more episodically, moving during high flows and then becoming locked in place, sometimes forming an armored surface that resists further erosion until an even larger event breaks it apart. The river therefore has a kind of memory: what happened during the last flood can determine how the bed responds to the next.

The floodplain, too, participates. During large floods, water spills over the banks, slowing dramatically as it spreads across vegetation and microtopography. Fine sediment carried in suspension settles out, draping the floodplain in silt and clay. Those fines can strengthen banks by forming cohesive layers; they can also nourish the very plants whose roots further bind the soil. The character of the corkscrew-driven meander is therefore not dictated by water alone, but by a partnership among flow, sediment, and living structure. A sandy, sparsely vegetated bank tends to yield easily, often producing broad, shifting bends or even multiple channels. A bank laced with roots or reinforced by cohesive mud can maintain a narrower channel and support stable, migrating meanders with steep cut banks.

There is a human-scale way to sense the corkscrew without instruments. In a canoe, enter a bend along the inside and you may feel the boat tug toward deeper water, then drift outward as the main current asserts itself. Near the outside bank, the surface can show streaks of foam or floating debris aligned with the strongest shear. Drop a weighted line and you may find the deepest point surprisingly close to the cut bank, not in the middle. If the water is clear enough, you can sometimes see sand ripples on the inner bar, their crests angled by the local flow. These are surface clues to a three-dimensional architecture.

A historical anecdote from the era of river steamships captures the practical consequences. Pilots on large meandering rivers learned to "read" the bend: the outside bank promised deep water but also dangerous snags from collapsing trees; the inside bar promised shallows that could ground a boat. In low water, a channel that had been safe the previous year might be blocked by a newly grown point bar; in high water, a new chute might tempt the unwary into a shortcut that would later close with sediment. Navigation was therefore a skill in applied geomorphology, learned not from textbooks but from repeated negotiation with a moving boundary between liquid and land.

Cutoffs, when they happen, are not merely geometric tidying. They are ecological and social events. An oxbow lake can become a nursery for fish, a refuge for amphibians, a trap for nutrients and contaminants. On working floodplains, a cutoff can strand a boat landing, change access routes, or alter groundwater levels in adjacent fields. The abandoned loop may continue to flood, but it floods differently: slower, quieter, more sediment-laden. Over time it can become a ribbon of wetland that stores water and carbon, a faint echo of the river's former insistence on curvature.

The most striking thing about a meander, once the mechanics are in view, is how the river seems to negotiate with itself. Each bend is a compromise between turning and going straight, between carving and building, between a path that fits the valley and a path that shortens it. The corkscrew flow is the mediator: an internal motion that translates curvature into work on the banks. From that translation comes the meander's signature geometry, and from that geometry comes a landscape where yesterday's channel is preserved as an oxbow, a scroll bar, a line of trees that prefer slightly drier ground, a memory stitched into the floodplain in curves.

5.2 The Architecture of Overload

A braided river can look like a mind changing its mind in public. From an airplane window it is a tangle of shining threads, splitting and rejoining around pale islands of gravel, then recombining only to fracture again. The pattern feels restless, but it is not random. It is what happens when a river is asked to do two incompatible jobs at once: carry water efficiently while also carrying *too much* sediment.

The word *braided* is literal. Instead of a single deep channel that curves, a braided river spreads its flow across multiple shallow channels separated by bars and islands. Those bars are not fixed banks; they are piles of the river's own burden, dumped when the water can no longer keep everything moving. The strange geometry is therefore an accounting problem made visible: the river's capacity to transport sediment has been outbid by the sediment supply arriving from upstream.

"Transport capacity" sounds technical, so it helps to translate it into an everyday experience. Imagine walking through a crowded train station carrying groceries. You can keep everything in your arms while you're moving steadily and have space to adjust your grip. Add more bags, or force yourself into a jostling crowd, and the same load becomes unstable. Things slip. You stop, rearrange, and perhaps set a bag down for a moment. A river behaves in the same spirit. The flow has a limited ability to keep grains rolling, hopping, or suspended. When the flow slows, shallows, or spreads out, grains are set down. When enough grains are set down, the bed rises, the flow becomes even shallower, and the setting-down accelerates. The river begins to *manufacture* obstacles to its own motion.

Engineers and geomorphologists have a compact way to express this bookkeeping, known as the Exner equation: a sediment-continuity relationship that links changes in the transported sediment flux to changes in bed elevation. The phrase is daunting, but the idea is as simple as a ledger. Consider a short reach of riverbed as a control volume. If more sediment enters that reach than leaves it, the excess must go somewhere, so the bed builds upward: *aggradation.* If more leaves than enters, the bed is mined and lowered: *degradation.* Many practical sediment models, including those used in river management software, phrase this in an almost conversational way: if the flow's transport capacity exceeds the available supply, the river erodes; if supply exceeds capacity, the river deposits. A braided river is a landscape-scale announcement that supply has won.

Supply can overwhelm capacity for several reasons, and in nature they often arrive together. A steep mountain catchment delivers fast water, but it also delivers abundant rock broken by frost, landslides, wildfire, and intense rain. Glaciers grind valleys into flour and dump it at their margins. Volcanoes blanket slopes with loose ash that storms can mobilize in a single season. Human activity can be just as effective: deforestation and poorly managed roads increase erosion; mining tailings and construction can inject sediment pulses; dams can starve one reach while forcing deposition in another. A braided channel is therefore not only a hydraulic state but a signature of the wider landscape's condition.

The most vivid braided rivers often appear where glaciers have recently done their work. An outwash plain below an ice margin is a factory for braiding: meltwater streams spill across a broad, low-relief surface made of poorly sorted gravel and sand that the glacier has dropped without much organization. Banks built from such material have little cohesion. They crumble easily, so channels can widen and shift laterally rather than being held in a narrow groove. The flow is also highly variable. Warm afternoons can increase meltwater; sudden rain on snow can produce a surge; cold snaps can reduce discharge. Each fluctuation rearranges the bars. The river becomes a set of temporary agreements between water and gravel.

A concrete example sits at the edge of the North American imagination: the braided rivers that drain the Alaska Range and other glaciated mountains. Many visitors first meet them from a highway bridge, where the water below is not one river but a wide, glinting fan, braids stitched across a bed of cobbles. The scene is beautiful and slightly alarming, because it feels as if the river might choose to occupy any part of its valley at any time. That is not far from the truth. When a thread of flow is blocked by a growing bar, it seeks a new path around it. When a new path captures more discharge, it scours deeper, becomes more stable for a while, and then is itself undermined by deposition upstream. The whole system moves like a living thing, but the logic is mechanical.

The most important mechanical step is the formation of bars. A bar begins as a patch where sediment momentarily accumulates: perhaps behind a large clast, perhaps where the channel widens slightly, perhaps where two threads meet and the combined flow cannot keep all the incoming bedload in motion. Once a slight bump exists, it changes the flow field. Water accelerates around it, scouring its flanks and depositing on its lee side. The bump becomes a bar with a distinct shape, often with a gently sloping upstream face and a steeper downstream slip face where grains avalanche. Bars can migrate, merge, split, and sometimes grow vegetated enough to behave like semi-permanent islands. In a braided system, bars are not exceptions; they are the framework.

Why does a river respond to overload by splitting rather than by simply building a higher bed in a single channel? The answer lies in how flow organizes itself when it becomes shallow and wide. A single, deep channel concentrates energy and can maintain a strong, coherent thalweg that sweeps sediment along. When the bed aggrades, the channel becomes shallower. For the same discharge, shallow flow has a larger fraction of its energy lost to friction at the bed. The flow is more easily deflected by small bed irregularities, and it is more prone to divide into separate threads. Each thread becomes a small conveyor belt, carving a local trough and depositing local bars. The river has, in effect, discovered a way to spread the work of moving sediment across multiple lanes.

That spreading increases resistance and dissipates energy. A braided pattern is therefore an "architecture" not in the sense of careful design, but in the sense of structural response: a particular way of arranging channels and bars that makes a given combination of water, slope, and sediment supply physically possible. The system can be understood as a compromise. The river cannot stay single-threaded without choking on its load, so it repeatedly invents new pathways around its own deposits.

A classic set of laboratory experiments by Stanley Schumm and Haroon Khan (1971) demonstrated this threshold-like behavior in miniature. In their flumes, changes in slope and sediment supply produced distinct shifts in channel pattern. At lower sediment loads or gentler slopes, channels tended toward single-thread forms. Push the system into higher sediment supply relative to transport ability, and it reorganized into multi-threaded, bar-dominated patterns. The experiments were not perfect copies of nature, but they captured something essential: braidedness is not a poetic label. It is a phase of river behavior that emerges when controlling variables cross a boundary.

Bank material plays a decisive role in where that boundary lies. Cohesive banks made of clay and stabilized by roots can hold a narrow channel together even under substantial sediment loads. Noncohesive banks made of sand and gravel are easily reworked, allowing channels to widen and split. This is why braided rivers are so common on fresh outwash and on alluvial fans emerging from steep mountain fronts. The material is loose; the river can rearrange its boundaries as readily as it rearranges its bed.

The human consequences of this instability can be serious. Braided rivers are notoriously difficult to bridge, protect, or confine. A levee built to restrain one thread may simply encourage another thread to grow elsewhere, sometimes against the very infrastructure meant to be safe. In gravel-bed braided rivers, the location of the deepest channel at low flow can migrate rapidly, complicating navigation and water intakes. During floods, a braided river can occupy a much larger portion of its valley floor, turning what looked like dry gravel into fast water within hours.

A historical anecdote from the age of railway expansion illustrates how quickly braided logic can collide with human plans. In many mountainous regions, early rail lines and roads were built along valley floors because gradients were gentle there. Engineers saw wide gravel plains as convenient corridors. Yet those plains were often braided riverbeds in disguise: apparently empty space that belonged, intermittently, to water. Accounts from nineteenth- and early twentieth-century projects in glaciated and mountainous terrain describe repeated washouts where channels shifted during snowmelt or storm floods, undermining embankments and carrying away bridge piers. The river was not "misbehaving"; it was doing the only thing available when asked to convey a seasonal surge of water plus a seasonal surge of gravel.

Overload is not only about the *amount* of sediment but also about its *grain size*. Coarse gravel requires more force to move than fine sand. A river can carry enormous quantities of silt in suspension while struggling to budge cobbles along its bed. Many braided rivers are gravel-bed rivers: they have enough power to move gravel during high flows, but not enough to keep it moving continuously. The result is episodic transport. Gravel moves in pulses during floods, then stops, building bars that persist through lower flows. Those bars then steer the subsequent floods, which means the river's memory can last from one season to the next.

That memory is complicated by a subtlety that modern sediment-transport research has emphasized: sediment movement is not always purely local. In the simplest picture, what happens to the bed at a point depends on the local difference between incoming and outgoing sediment flux. But real grains have "step lengths": distances they travel between periods of rest. When those step lengths are large compared to the scale of bedforms or even to the scale of a reach, deposition and erosion can be distributed in ways that feel non-intuitive. Work by Pelosi and Parker (2014), among others, formalizes how such nonlocal effects can change the shape of aggradational and degradational profiles. The popular translation is straightforward: sometimes the river does not deposit the extra gravel right where it first becomes overloaded; it can carry it a while, then drop it in a more diffuse pattern. In a braided river, this contributes to the sense that change happens "everywhere at once," as bars appear, grow, and migrate in response to sediment pulses that are not neatly tied to a single point upstream.

One of the most revealing ways to watch overload in action is to focus on a confluence where a sediment-rich tributary enters a clearer main stem. After a landslide or wildfire, a tributary can deliver a slug of sand and gravel that fans out at the mouth, forcing the main channel to widen and split around fresh deposits. The main stem may have been capable of carrying its usual load, but the sudden bonus overwhelms it, producing a bar complex that persists long after the initial event. River managers sometimes describe these as "sediment waves" moving downstream over years. The phrase is apt: not a wave of water, but a wave of raised bed and altered channel geometry, translating the history of a watershed disturbance into the language of shifting channels.

Braided rivers also invite a different way of thinking about "efficiency." A single channel seems efficient because it focuses flow. Braiding seems inefficient because it multiplies wetted perimeter and friction. Yet for an overloaded system, that inefficiency is the point. By splitting into multiple threads, the river increases friction and reduces the ability of any one thread to cut deeply. The system becomes wide, shallow, and mobile. That mobility allows it to rearrange bars continuously, creating local steepness and local acceleration where needed to mobilize grains. A braided river is therefore not a failed meander; it is a self-organized machinery designed, by physics, to handle a problematic budget.

For people living near braided rivers, the "design" can feel like a threat because it is indifferent to property lines and fixed infrastructure. Yet braided systems also provide gifts. Their wide gravel beds can recharge groundwater, filtering water through porous deposits. Their shifting channels create a mosaic of habitats: cold, fast riffles; quiet backwaters; newly exposed bars for nesting birds. In some places, braided plains are important corridors for wildlife precisely because they remain open and dynamic rather than becoming densely forested.

There is a final twist to the architecture of overload, one that links the restless braids to the broader floodplain. When deposition persists, the bed rises relative to the floodplain, and the river becomes perched, flowing along a ridge of its own making. On such aggrading landscapes, a different kind of channel change becomes more likely: *avulsion*, the abrupt abandonment of a channel in favor of a new course across the floodplain. Rudy Slingerland and Norman Smith (2004) emphasized how strongly avulsions are associated with aggrading floodplains: when the channel is elevated, it takes less of a breach for water to escape and carve a new path. Braiding and avulsion are not the same phenomenon, but both share the same underlying cause: too much sediment for the channel to comfortably carry. One expresses the overload within the channel belt as splitting and rejoining threads; the other expresses it across the floodplain as a wholesale relocation of the channel.

Stand on a gravel bar of a braided river at low flow and the system can appear almost serene. Water whispers through a nearby thread; sunlight flashes on ripples; the air smells of wet stone. The calm is real, but it is conditional. Somewhere upstream, a storm cell can turn a trickle into a brown surge, and the tranquil bar underfoot can become submerged, then reshaped, then erased. The architecture is built to be rebuilt. Its stability is not in fixed forms but in a reliable response to overload: when the ledger stops balancing, the channel does not merely adjust its speed. It rearranges its geometry until the burden can move again.

5.3 The Predictable Accident

Rivers have a talent for humiliating confident predictions. A levee is built to hold a channel; the channel gnaws at its toe. A map is printed; the blue line migrates. A neat theory is proposed; a flood arrives and redraws the argument in mud. Yet the humiliation is not total. Rivers change in ways that are locally surprising but statistically disciplined, like a crowd that seems chaotic at the level of shoulders and elbows but forms recognizable lanes at the level of the whole street. The question is not whether rivers are unpredictable, but *what* is predictable about them.

One of the most influential answers came from Luna Leopold and Gordon Wolman in the 1950s, at a moment when American rivers were being studied with an unusual mixture of curiosity and urgency. The postwar United States was building dams, expanding cities, straightening channels, and learning that a river's response to engineering can be perverse. Leopold and Wolman approached river patterns as an empirical problem: not "why does this particular bend exist?" but "under what physical conditions do rivers tend to adopt one pattern rather than another?" Their gift was to treat planforms—

meandering, braided, relatively straight—as alternative states of the same kind of system. The state a river occupies is constrained by a small set of controls, especially the slope of the land and the amount of water the channel regularly conveys.

The phrase *bankfull discharge* captures this second control. It refers to the flow that just fills a channel to the top of its banks before spilling onto the floodplain. Bankfull is not necessarily the biggest flood; in many rivers it is a moderately large event that happens every year or two. It matters because it is often the flow that does the most shaping work. Small flows lack the power to move much sediment or cut banks; very large floods may spread out and deposit more broadly. Bankfull is the frequent sculptor, the size of chisel a river reaches for most often.

Slope, meanwhile, is gravity's dial. All else equal, a steeper river has more potential energy to spend per unit distance. That energy can go into moving water fast, moving sediment, or eroding boundaries. Put slope and bankfull discharge together and you have a way to talk about river energy and river size without getting lost in details. Leopold and Wolman found that these two controls organize channel patterns in a surprisingly systematic way.

The core idea can be stated plainly: steep, sediment-laden systems tend to braid; gentler, lower-sediment systems with stronger banks tend to meander. That statement is not a moral about "wild" versus "tame" rivers. It is a statement about the balance between driving forces and resisting forces. A braided river is often a sign that sediment supply and slope push the channel toward wide, shallow flow that cannot maintain a single stable thread. A meandering river is often a sign that the channel can maintain coherence, allowing curvature-driven secondary circulation to sculpt cut banks and point bars.

Calling this a *predictable accident* may sound like a contradiction, but it fits the lived experience of rivers. The details of where a cutoff happens, or which braid thread will dominate after a flood, can be hard to foresee. The occurrence of cutoffs and channel switches is "accidental" in the sense that it depends on particular sequences of events: a log jam here, an ice jam there, an unusually timed flood. Yet the *kind* of river that produces frequent cutoffs, or frequent braiding, is predictable from its setting. The accidents are constrained by physics, and they repeat in recognizable ways.

A useful metaphor is handwriting. Two people may write the same word with slightly different loops and flourishes, depending on mood and speed. The exact wiggle of the pen is hard to predict. But the script—cursive versus block letters—is strongly constrained by training and by the design of the letters themselves. Rivers are like that. A particular bend is contingent; the tendency to meander is not.

This tension between contingency and constraint becomes clearer when thinking about meanders as a kind of instability rather than as a deliberate plan. In the 1980s, Alan Howard and Thomas Knutson (1984) demonstrated through simulation that a channel does not need a careful blueprint to become sinuous. Give the channel simple rules for lateral migration that depend on curvature, and allow cutoffs when bends become too tight, and complex meander patterns can self-organize from small perturbations. The details differ from one run to another, but the overall statistical character—wavelengths, distributions of curvature, the episodic creation of oxbows—emerges robustly. The river behaves as if it were improvising within a strict musical key.

A refinement of this idea came from work on how bend migration depends not only on local curvature but also on upstream geometry. Gary Parker and Edmund Andrews (2006) analyzed how the erosion rate at a point in a bend can be influenced by curvature upstream, because the flow takes a finite distance to adjust to changes in channel direction. The bank does not respond instantaneously to the bend beneath it; it responds to the bend as the flow *remembers* it over some length. This nonlocality matters because it ties the fate of one bank to the shape of the channel upstream. The river's accidents are therefore coupled. A small change in curvature can propagate its influence downstream, shaping where erosion will be strongest and where a cutoff might eventually be encouraged.

Nonlocality also helps explain why it is so difficult to "freeze" a meandering river in place without consequences. If the flow's erosive focus depends on upstream curvature, then armoring one bank in one location can shift the erosive energy elsewhere, sometimes with enhanced intensity. River engineers have learned, sometimes painfully, that local fixes can trigger remote responses. The predictable part is not the exact point of failure but the fact that the system will attempt to re-establish a coherent pattern of energy expenditure and sediment movement.

A historical anecdote makes this concrete. On many large rivers in the nineteenth and early twentieth centuries, navigation interests pushed for channel improvements: removing snags, closing side channels, and sometimes cutting off bends to shorten routes. On the Mississippi, engineered cutoffs reduced travel time, but they also altered slope locally by shortening the river. A shorter path over the same drop means a steeper gradient; steeper gradient can mean more erosive power. The river responded with bed degradation in some reaches and deposition in others, as it redistributed sediment to accommodate the new energy regime. The lesson is not that engineering is futile, but that it must negotiate with the river's tendency to seek a consistent relationship among slope, discharge, and sediment.

Leopold and Wolman's contribution shines in such moments because it provides a map of tendencies. Their slope-versus-discharge view does not predict the precise future of a single bend. It predicts the *neighborhood* of behavior a river is likely to occupy. In a low-slope, moderate-discharge setting with cohesive banks, a river pushed slightly out of alignment tends to return to a meandering mode, migrating and cutting off but maintaining a single-thread identity. In a steeper setting with abundant coarse sediment, attempts to impose a single deep channel may be undermined by bar formation and splitting, because the sediment ledger refuses to balance.

This is where an older, elegant idea about meanders also belongs: the notion that a meandering path might represent some kind of statistical or energetic optimum. Walter Langbein and Luna Leopold (1966) proposed a "minimum variance" framing in which meander geometry resembles a constrained random walk that minimizes changes in direction. The argument is subtle and has been debated, partly because rivers are not free to optimize without constraints imposed by bank strength, vegetation, and sediment supply. Yet the appeal of the approach remains: it suggests that the meander is not only a mechanical consequence of secondary circulation and bank erosion, but also a likely outcome when a noisy, perturbed system is forced to remain within certain limits. Even if one does not accept the idea of an explicit optimization goal, the broader message survives: the wandering is shaped by constraints that are themselves stable.

The phrase "channel pattern" can also mislead if it is taken too literally, as if meandering and braiding are rigid categories. In reality, many rivers occupy intermediate or mixed states. Some are "wandering" rivers, with a main channel that swings but also develops occasional bars and splits. Some switch between patterns depending on discharge, sediment supply, or human alterations. A river downstream of a dam may become more single-threaded as sediment supply is reduced; a river downstream of a wildfire or landslide may temporarily braid as sediment pulses arrive. Pattern is better thought of as a tendency, a preferred arrangement under a given set of controls, rather than a permanent identity.

The predictive power of Leopold and Wolman's framing is therefore not that it stamps a river with a single label for all time. It is that it encourages the right questions. What is the effective slope of the reach? What is the bankfull discharge, and how does it vary through the year? What is the sediment supply, and what sizes dominate? How strong are the banks, and what role do roots and cohesive mud play? When those questions are answered, the river's range of likely behaviors narrows dramatically.

There is also a human reason to care about predictability at this level. Floodplain societies often have a choice between treating river change as a series of isolated surprises or as a patterned process. Insurance, zoning, bridge design, habitat restoration, and dam operation all hinge on expectations about channel mobility. A meandering river's hazard is often lateral: bank erosion, migration, occasional cutoff. A braided river's hazard is often spatially diffuse: rapid channel shifting across a wide corridor, aggradation that can raise flood levels, sudden incision in one thread. Recognizing which regime a river tends toward is not merely an academic classification; it shapes which kinds of risk are sensible to plan for.

A concrete example can be found in rivers that drain steep, sediment-rich basins and then enter lower-gradient plains. Upstream, the channel may braid across an alluvial fan, repeatedly shifting position as coarse sediment is deposited. Downstream, where slope decreases and banks become finer and more cohesive, the channel may consolidate into a single thread and begin to meander, building a floodplain with scroll bars and oxbows. The transition is not a magic line but a zone where the sediment ledger changes: coarser grains drop out, finer grains continue, and the channel's ability to carve a stable path changes with them. One can often stand at such a transition and watch two different river logics on either side of a few kilometers.

The "accidents" of rivers become most dramatic when viewed at larger scales: avulsions that relocate a channel across its floodplain, abandoning settlements and fields. Slingerland and Smith (2004) emphasized that avulsions are strongly associated with aggrading floodplains, where the channel bed rises relative to its surroundings as sediment accumulates. Over decades to centuries, that rise makes the channel precarious: a breach during a flood can send water down a steeper path across the floodplain, and once a new route is established it can become the main channel. The timing of an avulsion can be hard to predict because it depends on flood sequences and local weaknesses. But the predisposition toward avulsion is predictable from the long-term sediment balance and the geometry of the floodplain. A river that is steadily building its bed is, in a sense, loading a spring.

The unsettling implication is that stability is sometimes an illusion created by short memory. A community may experience a century of gradual meander migration and assume the river is "behaving," only to face an avulsion that relocates the main channel in a matter of days. Conversely, a braided river may appear impossibly unruly until one notices the consistency of its response: every time sediment supply surges or discharge wanes, bars grow and channels split; every time a large flood arrives, the pattern is wiped clean and rewritten with familiar motifs.

Leopold and Wolman's enduring insight is that the river's motifs are not arbitrary. They arise because water under gravity must spend its energy while obeying the stubbornness of grains. When slope is steep and sediment is abundant, the channel spends energy by multiplying friction in a braided web. When slope is gentler and banks can hold, the channel spends energy by bending, using secondary circulation to exchange sediment between outer banks and inner bars.

Between these poles lies a continuum of forms, each one a workable compromise between what the landscape supplies and what the flow can remove.

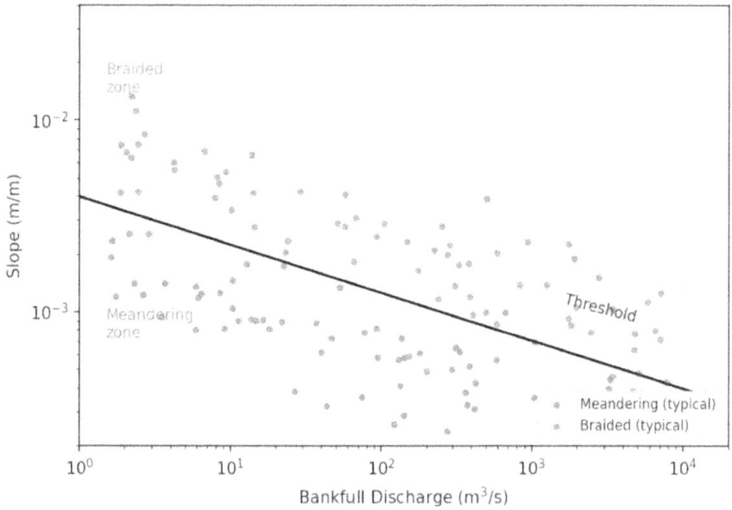

A plot like this is not a prophecy; it is a compass. Given a river's slope and the discharge that regularly fills its banks, one can often guess whether the channel will prefer to snake in a single thread or fracture into braids. The remaining uncertainty—the cut bank that fails tonight, the bar that emerges after next month's storm—belongs to the realm of local contingencies. Rivers keep their right to surprise, but they do so within boundaries that gravity, water, and sediment quietly enforce.

Chapter 6

The River's Lungs

A river confined solely to its channel is a system in distress, unable to perform its most vital regulatory functions. We step outside the banks to discover how the floodplain acts as a massive hydraulic capacitor, inhaling floodwaters to dissipate destructive energy and exhaling the nutrient-rich sediments that have supported civilizations for millennia.

6.1 Escaping the Straitjacket

Most of the time a river behaves like a disciplined machine: water funnels between banks, the current stays organized, and energy has a clear job to do. Gravity pulls downslope, friction drags back, and the channel negotiates a workable compromise. The river's force is easy to picture because it is concentrated. You can stand on a bridge and see it; the fastest thread runs dark and deep, surface ripples lean downstream, and anything buoyant snaps into that main conveyor belt.

A flood begins when that machine is asked to do more than its casing allows. Rain and snowmelt do not arrive as a polite trickle. They arrive as a pulse, sometimes sharp enough to feel like a blow. The channel responds first by speeding up and deepening. Then comes the moment when the river reaches the top of its banks and the geometry changes in an instant. The flow is no longer a ribbon; it becomes a sheet with a deep groove cut through it. The river has escaped its straitjacket.

That escape is not a simple spill. It is a radical shift in physics, like changing a musical instrument mid-note. Inside the channel, water has been forced into a relatively narrow cross-section. Outside it lies the floodplain: flatter, rougher, and full of obstacles that barely matter at low flow because they are not in the water's path. When water spreads laterally, three things happen together: the same volume of water is suddenly distributed across a much larger area; the boundary the water rubs against changes from mostly smooth bed and banks to a forest of stems, leaves, litter, fences, micro-topography, and buildings; and the flow has to share energy between forward motion and sideways motion as it negotiates that new space.

To make sense of the change, it helps to keep one everyday idea in mind: kinetic energy is the energy of motion. For a moving mass, it scales with the square of velocity. Double the speed and you quadruple the kinetic energy. Rivers are not just moving water; they are moving energy. The channel is a device for keeping that energy concentrated enough to transport sediment, erode banks, and carve patterns into the landscape. Overbank flow is what happens when the device opens and energy leaks into a much larger, messier world.

Hydraulics has its own language for the forces involved, but the heart of it is simple enough to picture at a dinner table. A river in its banks is like a crowd forced through a corridor: bodies align and move in the same direction, and the corridor walls keep them from spreading out. When the corridor opens into a wide room, the crowd fans out. People bump into tables and chairs; some stop entirely; others continue, but they can no longer all move at the same pace. Motion turns into jostling, detours, and pauses. In water, those detours and bumps are turbulence: swirling eddies that convert organized forward motion into internal friction and mixing. Turbulence is one of the river's most effective ways of dissipating energy, meaning it turns the river's ordered push into heat and chaotic motion too small to do geomorphic work at the scale of whole channels.

Within the channel, the fastest water tends to run above the deepest part, slightly away from the banks where friction is strongest. The velocity profile across the cross-section is shaped by shear, the gradient in speed from one layer to the next. Shear matters because it is a measure of stress: the river's grip on its boundaries and on the particles resting there. A fast current skimming over a bed of sand can pick grains up because the water exerts enough shear stress to overcome the grains' weight and the small interlocking forces between them. If the river slows, that stress falls quickly, and the grains drop out. Anyone who has watched a muddy floodwater creep across a field has seen the outcome written on the ground a day later: streaks of silt, little fans of sand, and a fine film that settles into every footprint and tire track.

The moment water leaves the banks, it encounters a new kind of shear: not only the shear between water and the ground, but shear between fast and slow water masses. The channel remains deep and comparatively efficient, so a core of fast flow continues along the river's former path. But in the overbank zone the water is shallow. Shallow flow has more contact with the ground per unit volume, which means more friction. Add vegetation and roughness and the floodplain can behave like a brake pad. The result is a sharp velocity contrast at the interface between the channel and the floodplain: a moving river next to a sluggish sheet.

That interface does not stay clean. The fast water pushes against the slow water and rolls into it; the slow water gets tugged and swirls back. Eddies peel away from the boundary like curls of smoke. This shear layer is a factory for turbulence. It steals energy from the organized channel flow and bleeds it into the floodplain in the form of chaotic mixing. To someone watching from above, the flood can look like a single body of water. Within it, the river is busily shredding its own momentum.

The floodplain adds another trick: it turns the river's forward energy into *storage*. To store water on a floodplain is not necessarily to trap it behind a wall; it can be as simple as letting it spread into hollows, around slight rises, and into the patchwork of low spots that every natural surface contains. Water pooled in a shallow depression has little velocity, so its kinetic energy is near zero even though the water still carries potential energy by virtue of being elevated above downstream reaches. In effect, the river trades speed for spread.

There is a paradox here that is easy to miss if floods are imagined as nothing but destructive overflow. Overbank flow can be violent at the leading edge, especially where it bursts through weak points or pours across man-made gaps, but the overall act of spreading is often an act of calming. A river that is allowed to occupy its floodplain can lose its sharpness. This is one reason floodplains have historically been both feared and coveted: they flood, but they also build and renew land.

The classic example sits at the beginning of many civilizations: the Nile's annual flood. For millennia, the Nile rose in summer as rains fell far upstream. The floodwater spilled onto the valley floor, slowed, and left behind fertile sediment before draining back. The rhythm became not just an agricultural schedule but a cultural fact. There were official measurements of river height, taxes tied to expected harvests, and entire administrative systems built around the pulse of water. The valley's prosperity was tied to this simple hydraulic transformation: once the river was free to spread, its kinetic energy was dissipated across a wide area, and its sediment load could settle. People did not need to know the equations for shear stress to learn, through experience, that still water makes fertile mud.

Even in places without a sacred flood calendar, the same physics shaped settlement. River towns tended to rise on natural levees, the slightly raised ridges built along channel margins. These ridges are made from the river's own behavior at bankfull and beyond. As floodwater spills out of the channel, it decelerates abruptly. The coarsest particles carried in suspension and as bedload cannot remain in motion under the reduced shear stress and drop out close to the bank. Finer silt and clay travel farther before settling. Over many floods, the result is a subtle topography: high ground near the channel and lower, wetter ground farther away. That high ground was a natural place to build: close to navigation and water, but slightly safer from the deepest inundation.

The mechanical reason those natural levees form is worth lingering on, because it shows how overbank flow is a sorting machine. Sediment transport is not all-or-nothing. Rivers carry a spectrum: gravel that rolls and hops near the bed; sand that can both bounce and suspend; silt and clay that can remain suspended for long distances. The ability to carry each fraction depends strongly on velocity and turbulence. When the river escapes, the system experiences an abrupt drop in carrying capacity, the maximum sediment load the flow can maintain without deposition. The drop is not uniform across the floodplain. Where flow remains active, such as along shallow flood channels or through openings in riparian vegetation, sand may continue to move. Where the water becomes nearly still, even fine silt can settle.

This is why floodplains record floods as patterns rather than blankets. The popular image is a uniform layer of fresh silt spread evenly like frosting. In reality, deposition can be patchy and sometimes surprisingly modest, while other processes, like the river's sideways migration over years and decades, contribute heavily to floodplain building. But the fundamental dissipation mechanism remains: as motion is broken up by roughness and storage, sediment that required speed to stay aloft loses its support and falls.

Roughness is the floodplain's underrated engineer. In a clean laboratory channel, roughness might be described by grain size on the bed. On a floodplain it becomes almost everything that sticks up: grass clumps, shrubs, tree trunks, fallen branches, crop rows, berms, the edges of roads. Each obstacle creates a wake, a region of slowed flow and swirling eddies. Multiply one wake by thousands of stems and you create a landscape-scale drag. Hydraulicians often roll this complexity into a single parameter, a roughness coefficient, but physically it is the same story repeated: the flood's kinetic energy is converted into turbulence and then into heat.

Vegetation adds another layer because it is flexible. A stand of reeds or tall grass does not behave like a set of rigid posts. At low depths it can stand upright and strongly resist flow, creating large drag and forcing water to back up. As depths increase, stems bend and align with the current, reducing drag. The floodplain's braking power is therefore not constant; it changes with stage, the height of the water surface. This helps explain why some floods seem to spread and stall, while others surge farther across the plain: the same field can be a sponge at one water level and a slick at another.

There is also the matter of air. In the channel, the dominant shear boundary is the bed and banks. Once the floodplain is inundated, a large part of the flow becomes shallow enough that the surface interacts more intimately with the atmosphere. Wind can push and wrinkle the surface; raindrops can dimple it; small waves can form. These are minor compared to bed friction, but they are part of the broader theme: the river's energy is no longer focused on carving one path; it is being shared with a broader set of interactions. The flood becomes a coupled system of water, plants, soil, and air.

One concrete way to see dissipation at work is to watch what happens to floating debris. In-bank, leaves and foam lines often shoot downstream along coherent streaks, tracing the fastest surface currents. During overbank flow, those streaks fray. Debris collects behind shrubs, along fence lines, and in the lee of trees. If you return after the water recedes, the debris marks the micro-hydraulics: a tangle of branches here indicates a local eddy; a long line of reeds pressed flat indicates a persistent jet of water; a patch of sand deposited in a swirl indicates a small region where turbulence slowed enough to let grains settle. The floodplain is, in this sense, a giant recorder of energy dissipation.

The straitjacket metaphor is more than poetry because confinement changes what the river must do with its energy. When lateral escape is blocked by high banks or artificial levees, the river cannot easily trade velocity for spread. The same inflow must be accommodated in a narrower space. If the cross-sectional area cannot expand sideways, it expands upward. Stage rises. With higher stage comes more hydraulic radius (a measure related to how much wetted boundary exists relative to flow area) and often higher velocity through a given reach, especially if the floodplain roughness is removed or bypassed. The river's energy remains concentrated, and the consequences travel downstream.

One historical example of the human struggle to keep the river in its casing is the lower Mississippi. Long before modern engineering, people built embankments to protect fields and settlements. As navigation, plantations, and urban centers grew, the levee lines lengthened and rose. The goal was understandable: keep the river off the land. But the mechanical bargain was harsh. The river that once spilled onto wide bottomlands was increasingly forced to pass floods within a limited corridor. During major events, water levels climbed higher against the walls, and the pressure to breach became immense.

When breaches occurred, they were not gentle overbank spreads but explosive releases, because the stored water in the channel and its elevated stage suddenly found a low, weak point. The escaping jet could scour deep crevasses and carve new channels across the floodplain, a reminder that denying a river its gradual pressure release can invite a more violent one.

The difference between a controlled overbank flood and a levee breach is, again, a difference in energy pathways. In a gradual overtopping, the river begins to share flow with the floodplain early. The interface shear layer grows, roughness begins to do its work, and the flood wave's peak can be shaved. In a breach, the system behaves more like puncturing a pressurized container: the local velocity through the opening can be enormous, and the energy is delivered as an erosive blast rather than a distributed sheet. The floodplain still dissipates energy, but it does so after the river has already spent a large portion of its power on destruction.

It is tempting to treat floodplains as passive basins, but they have structure. Many floodplains contain secondary channels, swales, and old meander scars that guide water during floods. A meander cutoff may leave behind an oxbow lake; even when it fills with vegetation, its slight depression can still route floodwater. These features act like veins in the floodplain's body, creating preferential pathways where flow remains active rather than stagnant. From the standpoint of dissipation, they create a patchwork: some zones strongly attenuate motion and trap sediment; others carry water more efficiently and can even concentrate it in unexpected places.

This patchiness makes floodplains simultaneously protective and unpredictable. For a farmer, a small depression that quietly fills every spring can be a nuisance but also a reliable source of moisture and fertility. For an engineer, that same depression can become a corridor that delivers floodwater toward a road embankment or a neighborhood, because water follows the easiest path across a flat landscape. The underlying physics stays the same, but the outcome depends on the floodplain's micro-topography and roughness.

Another subtle mechanism of energy loss arises when floodwater infiltrates the ground. Floodplains often have permeable soils near the surface, built from layers of sand and silt laid down in past events. When water spreads and slows, it has time to soak in. Infiltration does not remove much volume during the peak of a very large flood, but it is a quiet form of dissipation: water that enters pores loses kinetic energy and becomes part of the subsurface flow system, moving slowly under different rules. It also changes the floodplain's future behavior by altering soil moisture, affecting vegetation, and influencing how much rain can be absorbed in the next storm.

A final piece of the mechanics concerns how floods end. When the water level in the channel drops, floodplain water does not always retreat at the same pace. Because the floodplain is broad and rough, and because it may have shallow depressions cut off from efficient return paths, water can linger. This creates a long tail: the river's quick, forceful pulse becomes a slow release. Even without invoking any circuitry metaphors, the physical picture is clear. A river that can escape can also exhale. It does not simply dump excess water downstream; it spreads it, slows it, and then gathers it back over time.

Seen from the scale of minutes, overbank flow can look like chaos: water spilling in every direction, swirls around trees, ripples crossing ripples. Seen from the scale of decades, it is an ordering process. The floodplain is shaped by the repeated conversion of speed into spread, of momentum into turbulence, and of suspended loads into settled layers. The river is not losing energy in a moral sense, as if becoming weaker; it is spending energy to do a different kind of work. Instead of deepening a single groove, it performs maintenance on its wider valley: building natural levees grain by grain, filling old channels, refreshing soils, and redistributing the materials that make landscapes habitable.

A river confined to its channel is like breath held too tightly: efficient, focused, and potentially dangerous when the pressure rises. A river allowed to spill has room to dissipate its force. The water still moves inexorably downhill, but it does so with a changed posture, trading the sharp authority of a narrow current for the broad, friction-rich touch of a flood across land. When the flood recedes, what remains is not only a line on a gauge or a mark on a bridge pier, but the quiet evidence of energy transformed: a thin skin of silt on leaves, a bar of sand at the edge of a field, a tangle of reeds laid flat where the current leaned hardest, and a valley floor subtly rearranged by the simple act of giving the river space.

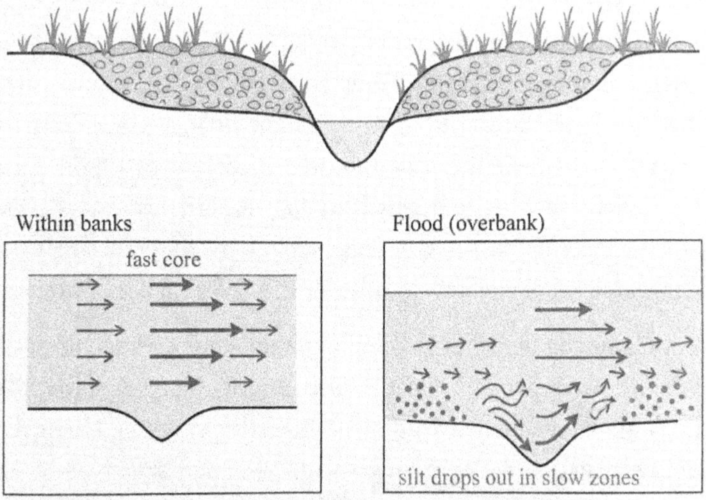

Velocity concentrates in the deep channel; once water spreads onto the floodplain, roughness and shear produce **deceleration, turbulence, and deposition.**

6.2 Hydraulic Capacitors

A flood is often described as a volume problem: too much water arriving too quickly. That is true, but incomplete in the way a medical chart can be true and still fail to capture what it feels like to be ill. Rivers do not merely *contain* water; they *transmit* it. A storm in the hills is converted into a moving signal, a pulse that travels down the branching network. Along the way, that pulse is stretched, clipped, delayed, split into side paths, recombined, and sometimes sharpened into something more dangerous than what entered upstream. The landscape is not just a backdrop; it is a signal processor.

The simplest instrument for seeing that processing is a hydrograph: a plot of discharge (how much water passes a cross-section per unit time) against time. Hydrographs are the river's heartbeat tracings. A steep rising limb means water is arriving quickly; a high peak means the channel is being asked to carry a large flow; a long falling limb means the catchment is releasing stored water slowly. Engineers and hydrologists treat hydrographs as data, but they also carry a story that anyone living near a river understands: the difference between a manageable swell and a wall of water is often a matter of timing.

Floodplains change timing by giving floods somewhere to go besides straight downstream. A river confined to its banks behaves like a wire: it transmits the pulse efficiently. A river connected to its floodplain behaves more like a device with memory. When water spills laterally, some of it keeps moving, some of it slows to a crawl, and some of it becomes, for practical purposes, stored. The stored portion is not destroyed; it is *borrowed* from the flood's peak and repaid later. That borrowing-and-repayment is why the floodplain works like a capacitor in an electrical circuit.

A capacitor, in everyday terms, is a component that stores charge when voltage rises and releases it when voltage falls. It smooths spikes. It turns a jagged input into a calmer output, often at the price of a delayed response. Swap charge for water and voltage for water-surface slope, and the analogy becomes physically useful rather than merely cute. When the river's stage rises quickly, water spills into side storage and the main channel peak is shaved. When the stage begins to fall, that stored water drains back, sustaining flow longer than it would otherwise persist. The flood wave becomes lower and broader. The river, in effect, exhales slowly instead of coughing sharply.

Hydraulic models make this mechanism visible in a way that field observation alone often cannot, because the key action is the exchange of water between fast and slow zones. In the U.S. Army Corps of Engineers' HEC-RAS modeling system, one practical way of representing the difference is to split a cross-section into *active flow* and *storage*. Active flow is water moving with appreciable velocity; storage is water that occupies space but has little velocity, like water ponded behind a natural levee or in a backswamp. When a floodplain is connected, a larger fraction of the cross-sectional area can behave as storage at high stage. That storage does not need to look like a lake. It can be a sheet of shallow water, slowed by roughness and blocked by micro-topography, that nonetheless holds a substantial volume when spread over kilometers.

The physical bookkeeping is straightforward. A flood wave carries a certain volume. If some of that volume is temporarily held outside the fast conveyor belt of the channel, the channel's peak discharge must drop, because discharge is volume per time. The only way around that arithmetic would be if the stored volume were filled and emptied so quickly that it effectively remained part of the main through-flow. That caveat matters, and it is one reason floodplains can behave in surprising ways.

It helps to picture a storm rolling across a basin, delivering heavy rain for several hours. In the headwaters, small channels respond quickly: water rises, peaks, and falls in a short time. Downstream, tributaries contribute their own pulses. When peaks from different branches arrive at the same time, they add constructively, like synchronized applause becoming deafening. When they arrive out of phase, they partially cancel, like overlapping waves on a pond. Floodplains influence both processes. They can delay the passage of a peak along a main stem, changing its alignment with tributary peaks. They can also attenuate the peak so that even if timing aligns, the combined crest is less extreme.

This is not abstract. People have watched rivers become more dangerous not simply because rainfall increased, but because storage and delay were removed. Channel straightening, levees, and drainage ditches speed the delivery of water to the main channel. Wetlands are drained; small depressions are tiled; tributaries are turned into efficient gutters. The basin's pulse becomes sharper. A sharper pulse demands a higher peak discharge to pass the same volume. Downstream residents experience that as a river that rises faster, crests higher, and gives less time to respond.

One of the most vivid modern demonstrations of how timing and storage matter came during the Mississippi River flood of 1993. Weeks of persistent rainfall across the Midwest saturated soils and fed a long, rolling sequence of rises. The disaster was not a single flash flood but a prolonged siege in which tributaries kept delivering pulses into already swollen main channels. In many places, levee systems protected local areas until they didn't, and then failure could be abrupt. The event imprinted a lesson that goes beyond the drama of breached earthworks: when a river network is pushed near capacity for a long time, the landscape's ability to temporarily store water becomes the difference between a peak that stays within bounds and one that exceeds them.

The capacitor idea, though, can mislead if it is treated as a slogan: *more floodplain connection equals less flooding.* In hydraulics, the details of connection are everything. A floodplain is not a single bathtub; it is a patchwork of pathways and pockets. Some parts fill easily through low points in natural levees or through secondary channels. Other parts only inundate during larger floods. Some drain back quickly through efficient return channels. Others hold water for days or weeks, especially if the floodplain sits behind subtle ridges that trap water. These differences determine whether storage is accessed at the right moment to clip the peak.

Modeling work by N. Tull and P. Passalacqua (2025) captures this nuance by separating two limits on attenuation. Sometimes the limiting factor is *connectivity*: the floodplain has plenty of potential storage, but the flood wave cannot access it because the openings are too few, too high, or too constricted. In that case, increasing connectivity (for example, opening side channels or removing barriers) can dramatically increase attenuation because it allows water to spill out earlier, stealing volume from the rising limb when it matters most. At other times the limiting factor is *storage*: the floodplain is well connected, but it fills up. Once full, additional inflow cannot be stored, so the peak passes downstream largely unmodified.

Secondary channels make the story more interesting still. They can act like additional valves that open earlier, routing water onto the floodplain at moderate discharges and spreading it across a larger area. That sounds unambiguously beneficial, but there is a twist: if those channels also provide efficient drainage, they can shorten the time water remains stored. During very large floods, efficient pathways can allow the floodplain to fill and then return flow quickly, potentially reducing attenuation and, in some configurations, even sustaining higher downstream flows at the time of peak. The capacitor still exists, but its capacitance and its leakage rate change with stage.

This is one reason arguments about floodplain reconnection can become emotionally charged. The physics does not hand out simple guarantees. A connected floodplain can reduce peak flows in many situations, and it often reduces *water levels* locally by giving water more cross-sectional space, but the network response depends on where storage is located, how quickly it fills, and how it drains. A reconnection project that behaves like a generous storage basin for a two-year flood may behave like a quickly filled, quickly drained bypass during a rarer event. The same patch of land can be a safety valve in one year and a fast lane in another.

Floodplain roughness, introduced earlier as a brake on velocity, also matters here because it changes residence time. If water spreads across a smooth surface, it can keep moving as a broad, shallow current, staying part of the downstream signal. If it spreads into a rough forested bottomland, it slows, swirls, and ponds, shifting from conveyance to storage. In practice, two floodplains of the same size can have very different hydraulic personalities depending on vegetation, micro-topography, and the presence of old channels that act as sluiceways.

These effects show up clearly in the shape of hydrographs. Consider an upstream reach that has been leveed and straightened, and a downstream reach where the river still has access to its floodplain. The upstream hydrograph might rise quickly to a narrow, high peak and then fall quickly once rainfall ends. Downstream, the hydrograph may show a lower crest and a longer recession, reflecting water that was temporarily stored and then released. The river has not been made smaller; it has been made slower.

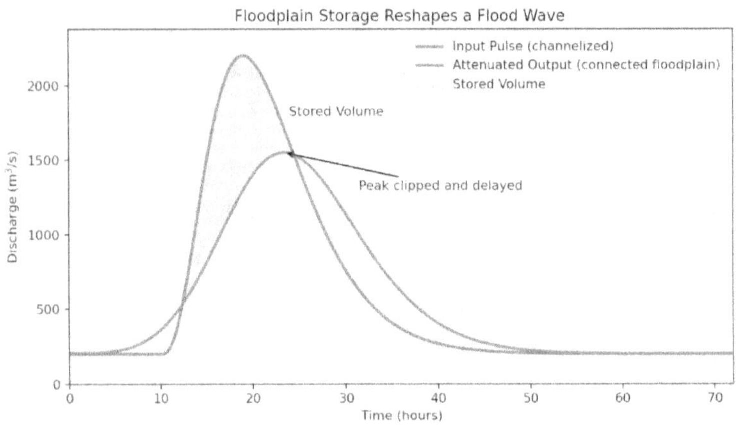

The shaded area in such a diagram is not just a graphical convenience. It represents a real volume, the water that is temporarily not in the main downstream-moving thread. In a real floodplain, that volume may be spread thinly over a vast area, so that no single spot looks dramatically deep. That thinness is part of the point: shallow storage is hydraulically powerful because it increases wetted area and roughness, which slows and holds water without requiring steep banks.

The capacitor analogy also clarifies a common misunderstanding about "flood control." Many people imagine flood management as a binary choice between stopping water and letting it through. Floodplains operate in a third mode: *they let water through by delaying it.* Delay can be as valuable as reduction, because it changes coincidence. A downstream town may not need the peak to be reduced to zero; it may need the peak to arrive six hours later, when a tributary crest has already passed. In a branching network, the most dangerous floods are often produced not by the largest individual pulse but by the alignment of several moderately large pulses. Storage that desynchronizes those pulses can reduce extremes without changing the total volume delivered over days.

This is one reason "room for the river" policies, wherever they are tried, often emphasize space rather than walls. A setback levee, for example, does not abandon flood protection; it relocates it, trading a narrow, high-energy corridor for a wider one that can accommodate storage and roughness. In the Netherlands, where water management is a matter of national identity, projects have included lowering floodplains, moving dikes inland, and creating side channels. The point is not romantic restoration. It is signal shaping: reducing the amplitude and steepness of the flood wave and buying time.

Yet capacitors can be overwhelmed. Any capacitor has a maximum charge it can hold. Any floodplain has a maximum volume it can store before it becomes simply another part of the moving flood. During extreme floods, storage fills, and the system shifts regimes. The river then behaves more like a wide channel than a channel-with-storage. At that stage, the floodplain still provides conveyance and can reduce local stage by increasing flow area, but its ability to further attenuate the peak can diminish. The hydrograph can still broaden, but the crest may not drop much more, and the downstream benefit may be smaller than intuition expects.

There is also a human dimension that complicates the clean picture of storage and release. People often build on floodplains precisely because they are flat, fertile, and close to water. Once built upon, the floodplain's roughness changes. Houses, roads, and walls can create new obstructions, sometimes increasing local storage but also creating dangerous flow concentrations. Drainage infrastructure can increase leakage, turning long-lived storage into faster return flow. The capacitor acquires new resistors and short-circuits. The network signal changes again.

The most honest way to speak about floodplains, then, is not as a guaranteed shield but as a physical opportunity. They are places where water can trade speed for space, where the sharp edges of a hydrograph can be softened, where a pulse can be made less violent by being made less sudden. That opportunity is carved into the geometry of valleys and written into the very idea of a floodplain: a plain that floods, not as an accident, but as a recurring function.

When water spreads out and slows down, a different clock begins to govern the river. Instead of minutes and meters per second, the relevant units become hours and centimeters per hour: seepage into soil, slow drift through forest understory, gradual return through back channels. The river's message still arrives downstream, but it arrives as a long sentence rather than a shout, carrying the same meaning with less damage.

6.3 The Silt Paradox

Floodplains are magnets. Stand on one and it is easy to understand why: the ground is flat enough to build on, close enough to water to irrigate, and often rich enough to grow crops that would struggle on thin upland soils. Many of the world's great cities and breadbaskets sit on these broad valley floors. The attraction is so strong that it can feel like common sense, as if rivers naturally come with free, fertile land attached.

The paradox is that the very qualities that make floodplains desirable are created by the thing people fear most: flooding. Floodplains are not simply *next to* rivers. They are made, maintained, and periodically renewed by the river's tendency to escape its banks. When societies decide that flooding is intolerable, they often engineer the river into a narrower corridor using levees, revetments, and straightened channels. That seems like a clean solution: keep the water in the channel and protect the land. Yet, in many settings, disconnecting a river from its floodplain changes the river's physics in ways that raise water levels, move risk elsewhere, and starve the valley of the sediment and nutrients that built it. The land remains alluring, but the bargain becomes unstable.

The silt paradox has a second layer, quieter and more scientific. Folk memory and schoolbook lore often treat floodplains as if they were built by generous, uniform blankets of silt laid down year after year. There is truth in that story: floods do deposit fine sediment, and those deposits can be agriculturally transformative. But detailed measurements and long-term observations show that floodplain construction is not always a steady rain of mud. In many rivers, the average thickness deposited in a large flood can be surprisingly small, while other processes—especially the river's sideways migration and the slow rearrangement of channels—do much of the heavy lifting. Floodplains are not only sediment traps; they are moving mosaics of erosion and deposition. That matters because it changes how one should think about "saving" floodplains with levees: the floodplain is not just a victim of water but also a product of it.

Begin with the simplest picture: water rises, spills out, slows down, and drops sediment. That is the image most people carry, and it is not wrong. It is also incomplete because it hides how selective the river is. A flood does not drop "silt" in general; it sorts a mixed load. Coarser grains fall first, close to where the water escapes. Finer particles travel farther, sometimes all the way to backswamps and oxbow lakes. Organic matter and nutrients hitchhike on that mineral load, especially in fine fractions, because clay particles and decomposed plant material have large surface areas that bind phosphorus and carry nitrogen in complex forms. The floodplain is therefore not only a depositional surface but also a chemical filter, an enormous contact zone between water and land.

This is why the world's classic floodplain civilizations did not merely endure floods; they used them. Along the Nile, the annual flood historically brought not just water but suspended sediment that refreshed fields. Along the Yellow River, floods were both fertilizing and catastrophic, a volatile mixture that shaped governance as much as agriculture. Along the Mississippi, bottomland forests and alluvial farms prospered in soils built from repeated inundation. The same material that makes a floodplain a breadbasket can make it a hazard zone. The paradox is not philosophical; it is mechanical. Fertility and danger are generated by the same pulse of water.

In the nineteenth and twentieth centuries, many societies attempted to resolve the paradox by forcing rivers into more predictable shapes. Levees became higher and more continuous. Channels were straightened or armored. Floodplains were drained for agriculture and development, with networks of ditches and pumps designed to remove standing water quickly. Each intervention had its own local logic. Together, they altered the stage—discharge relationship: the link between how much water a river carries (discharge) and how high it rises (stage). If a river is given less room to spread laterally, the same discharge often corresponds to a higher stage. Imagine squeezing a flexible tube: the volume still has to pass, so the water surface rises. Even when velocity increases and the channel scours, the confinement can produce a river that sits higher against its banks, like a load-bearing beam raised on temporary supports.

This is one aspect of the "levee effect": levees can encourage development behind them because they create an impression of safety, and that development increases the consequences when a flood exceeds design assumptions. There is also a hydraulic effect: confinement can speed the downstream propagation of flood waves by reducing storage and roughness, shifting flood timing and sometimes increasing peaks farther down the system. None of this means levees are foolish; many protect lives and critical infrastructure. It does mean that levees are not a simple subtraction of risk. They are a redistribution of water's work.

Sediment is where the redistribution becomes strangely literal. When floodplains are connected, some of the river's sediment load leaves the channel during floods and settles on the floodplain. This can be viewed as a loss to navigation channels, because sediment that settles can reduce channel depth if it deposits in the wrong place. It can also be viewed as a gain to the landscape, because sediment deposition builds soil, raises land relative to water level, and transports nutrients onto the valley floor. Disconnect the floodplain, and that sediment must go elsewhere. Some will deposit within the channel, forcing dredging or raising the bed (aggradation). Some will be carried downstream, potentially increasing sedimentation in reservoirs, deltas, and coastal waters, or, if trapped upstream, depriving those same coasts of their building material. The sediment budget—the accounting of sources, transport, storage, and export—changes everywhere at once.

A striking example of sediment interruption is the Nile after the completion of the Aswan High Dam in the 1960s. The dam delivered clear benefits: it smoothed seasonal flow, enabled perennial irrigation, and provided hydroelectric power. But it also trapped a large fraction of the sediment that once moved downstream. The Nile Delta, already subject to subsidence and global sea-level rise, became more vulnerable to coastal erosion when the river's sediment supply was reduced. The shoreline does not care why sediment is missing. It responds to the balance between incoming material and the sea's capacity to remove it. A delta is land made from sediment piled at the edge of the ocean; starve it, and it can shrink.

The Nile case is often told as a cautionary tale: trap sediment upstream, lose land downstream. The truth is more nuanced and therefore more instructive. Delta erosion is multi-causal. Sea-level rise, land subsidence from natural compaction and human groundwater extraction, and coastal currents all play roles. The dam is not a single villain but an important alteration in a tightly coupled system. The broader lesson is that sediment is not waste. It is a form of capital, accumulated slowly and spent quickly when the supply line is cut.

That same logic applies inside river valleys. When a river is cut off from its floodplain, the valley can experience sediment starvation in the places that once received overbank deposits. The land may remain fertile for a time—floodplain soils are not emptied overnight—but the long-term renewal slows. Meanwhile, the river continues to move sediment within its confined corridor. If the bed rises, stage for a given discharge can rise as well, increasing pressure on levees. If the bed scours instead, the channel may deepen, which sounds beneficial until it undercuts banks and infrastructure, lowers adjacent groundwater tables, and changes habitat. Either way, the river is still doing geomorphic work; it is simply doing it in a narrower zone with less opportunity to spread consequences.

The paradox becomes sharper when nutrient retention enters the picture. Floodplains are not just places where sediment settles; they are biogeochemical reactors. When floodwater spreads across vegetated land, it slows and lingers. That residence time allows particles to settle and allows dissolved nutrients to interact with soils and plants. Nitrogen can be transformed by microbes; phosphorus can sorb onto mineral surfaces; organic matter can be buried and stored. Researchers such as G. Noe and C. Hupp (2009) have shown that, at watershed scales in coastal plain settings, connected floodplains can retain substantial fractions of annual sediment and nutrient loads. That means a floodplain is not merely a local feature; it can influence downstream water quality and even coastal eutrophication.

Yet even here, the story refuses to be simple. Floodplains can also export sediment and nutrients when their banks erode or when stored material is remobilized in later events. A deposit is only "retained" if it stays. Hopkins and Noe (2018) emphasized the importance of net accounting: deposition minus erosion. A floodplain that quietly accumulates silt in its interior can simultaneously lose material at its margins as the river eats away at a bend. In some cases, bank erosion can overwhelm overbank deposition so that the floodplain, taken as a whole, is a net exporter of sediment even while it remains a net sink for certain nutrients. Phosphorus tends to bind strongly to particles; nitrogen cycles through more mobile forms. The time scales can differ too: phosphorus may remain stored for long periods, while nitrogen can be transformed and released on shorter biological clocks.

This helps explain why floodplain restoration projects can yield mixed results depending on what is measured and over what period.

A reconnected floodplain may trap sediment during moderate floods, improving downstream clarity, but it may also experience bank erosion during high flows that releases pulses of sediment. From the viewpoint of a downstream water treatment plant, those pulses matter. From the viewpoint of long-term soil building, the slow accretion matters. From the viewpoint of ecology, the pattern of inundation—the hydroperiod, the duration and frequency of flooding—can matter more than total annual sedimentation, because it controls plant communities and microbial processes.

Concrete evidence of reconnection's measurable effects can be found in unplanned experiments: levee breaches. In Iowa, for example, a levee breach along the Maquoketa River in 2010 reconnected parts of the floodplain, and subsequent monitoring over multiple years allowed scientists to quantify changes in sediment and nutrient deposition and movement. Such events are tragic for landowners, but they reveal how quickly a floodplain can resume its old role when water is allowed back onto it. They also reveal how much modern landscapes depend on artificial disconnection: it often takes a dramatic failure for the river to reach land that once flooded routinely.

The silt paradox is therefore not only about a clash between human desire and natural behavior. It is about competing definitions of "good." For a farmer, silt can be fertility. For a reservoir manager, silt can be lost storage capacity. For a navigation authority, silt can be dredging costs. For a delta community, silt can be survival. A single grain of clay can be simultaneously a nutrient carrier, a pollutant vehicle, and a building block of land. Rivers move that ambiguity downstream.

There is also a temporal paradox that plays out in slow motion. The benefits of floodplain deposition accrue quietly and cumulatively. A millimeter of mud added each year is barely noticeable, like a calendar page turning. The costs of a major flood arrive in a day. Human attention is naturally tuned to disasters more than to maintenance.

That bias pushes societies toward building higher walls rather than preserving slow, distributed processes. It is easier to celebrate a levee's completion than to celebrate a floodplain that, for the thousandth time, took a little bite out of a peak, trapped a little phosphorus, and raised itself a fraction of a millimeter. Yet over centuries, those fractions decide whether land stays above water.

One can see the consequences of ignoring sediment maintenance in the world's sinking deltas. Louisiana's coastal land loss is often discussed in terms of hurricanes and sea-level rise, but it is also a story about sediment delivery being interrupted and rerouted. When a river is prevented from flooding its delta plain, the plain subsides while sea level rises, and the land must be kept artificially dry or it becomes marsh and open water. The river's sediment, once spread widely, is sent down a deep channel into the Gulf or trapped behind control structures. The delta becomes a museum piece: a landform built by a process that has been largely turned off.

This brings the paradox back to the floodplain beneath one's feet. If the river is the valley's builder, then a levee is not just a wall; it is a decision about where building is allowed to happen. The wall protects the interior from inundation, but it also denies the interior new material and new chemical inputs that once arrived for free. Over time, land behind levees can become a kind of aging infrastructure, dependent on pumps, drainage ditches, and maintenance, while land outside the walls continues to be reshaped by deposition and erosion. The river does not stop working. It works where it is permitted.

None of this implies a romantic return to uncontrolled flooding. People live on floodplains because floodplains are good places to live, until they are not. The challenge is to hold two truths at once: floods are hazards, and floods are also the mechanism by which floodplains remain fertile, elevated, and chemically active. When those truths are separated—when floods are treated only as enemies—the landscape's hidden services can be lost, and the remaining risk can become sharper rather than smaller.

A useful way to think about silt is as the river's breath made visible. During high water, the river inhales from hillslopes and banks: soil particles, organic fragments, dissolved nutrients. When water spreads and slows, the river exhales onto the floodplain: a film of fine sediment, a scatter of seeds, a pulse of nutrients that will be taken up, transformed, or stored. Cut the river off from its floodplain, and the breath has nowhere to go. The river keeps inhaling, but it exhales downstream, often into places that cannot use the gift and suffer the burden instead.

In the aftermath of a flood, when the water has retreated, the floodplain often looks deceptively ordinary. Mud dries into a dull crust. Leaves are matted against fence lines. A faint stain marks how high the water rose. These are small signs, easy to dismiss in the urgency of cleanup. Yet they are also the physical record of a vast exchange: volume temporarily stored, energy dissipated, sediment sorted and redistributed, nutrients shifted between aquatic and terrestrial worlds. The floodplain's fertility is not a trait it *has*; it is a trait it *receives*, again and again, as long as the river is allowed to spill and return.

Chapter 7

Building the Edge of the World

When a river meets the sea, the logic of the catchment is inverted: the water stops gathering and begins to scatter, shedding its load to construct new land against the eroding tides. This chapter examines the paradox of the delta-a landscape that is simultaneously built by the river's sheer power and doomed by its own weight.

7.1 The Great Unraveling

A river network is usually easy to read from a map. Thin blue lines gather into thicker ones, like threads twisted into rope. Up in the headwaters, water is a rumor: a seep under moss, a trickle between stones, a ditch by a road. Downstream, it becomes insistence. Tributaries join, the channel enlarges, and the river's logic seems almost moral: many small contributions becoming one strong current.

Then, at the edge of the world the river has built for itself, that logic flips.

The flip is so common that it can feel inevitable, but it is not a matter of geography alone. Something physical happens at the river mouth, something that turns convergence into divergence. The river runs out of slope, meets a body of standing water, and abruptly loses the ability to carry what it has been carrying. The river, which spent its long descent sorting and selecting grains as if it had preferences, reaches the sea and becomes strangely clumsy, dropping sediment in its own path like a traveler unloading a pack in a doorway.

The result is a distributary network: not a tree whose branches gather inward, but a fan whose branches spread outward. On a satellite image it can look like a living thing opening its fingers into the ocean. On the ground it can feel like confusion made permanent: channels that split and split again, islands forming, disappearing, reappearing somewhere else; land that is neither quite land nor quite water, but a shifting compromise between the two.

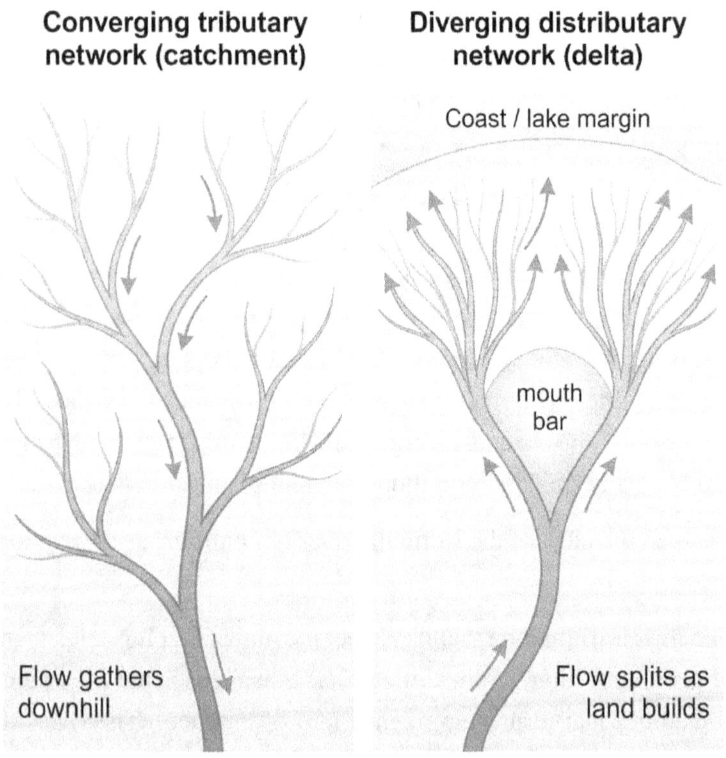

The diagram makes the reversal look tidy. Real deltas are not tidy. Yet the underlying switch in network topology is real: a tributary system is a funnel, while a distributary system is a sprinkler. The reason is not that the river suddenly wants to branch. The reason is that the river loses speed, and losing speed changes everything.

Speed is a river's carrying power. It is not exactly the same as discharge (the volume of water moving per unit time), though the two are often linked. A river can have a high discharge and still slow down if its channel widens and deepens, or if it enters a region where the slope flattens. What matters for sediment is the stress at the bed and the turbulence in the flow. Those determine whether grains are picked up, rolled, bounced, or kept in suspension. Upstream, the river is selective: coarse grains travel as bedload, moving in fits and starts; finer grains can be held aloft for long distances. The mix depends on the river's energy budget, which comes from gravity acting down a slope.

At a river mouth, the slope abruptly approaches zero. The river does not stop, but it has to reorganize the way it spends its energy. Imagine a fast jet of water leaving a hose into a calm swimming pool. As soon as the jet meets the pool, it spreads and slows; its edges curl into the quiet water, and the flow begins to mix and fan out. A river entering a lake or the sea is a larger, messier version of that jet. Engineers call it a *free jet*: the flow is no longer constrained by banks on both sides. The channel that guided the river becomes, at the mouth, a suggestion rather than a prison.

This is where the unraveling begins. The jet decelerates rapidly because it is pushing into water that is already there. In a lake, the receiving water may be fresh like the river, making mixing relatively easy. In the ocean, the receiving water is salty and denser; the river's fresh water tends to ride over it at first, like oil over vinegar, though turbulence quickly complicates that picture. Either way, the key is that the jet spreads: what had been focused into a narrow channel becomes distributed over a broader front. The same discharge now occupies a larger cross-sectional area. Velocity drops. Turbulence changes character. The river's capacity to keep sediment moving drops with it.

Sediment responds immediately and with little sentimentality. Coarse sand that could be nudged along the bed in the confined channel is now asked to move across a widening, slowing, shallowing flow. It cannot, so it settles. Finer particles may travel farther, sometimes far enough to color the water a milky brown offshore, but even they are affected by the loss of velocity and by changes in turbulence. What matters most for the birth of a delta is that deposition begins right where the river would prefer it not to: at the mouth itself.

The first major deposit is often a *mouth bar*. The name sounds like a place to order a drink, but it is a kind of sandbank that forms just offshore or just inside the mouth, where the decelerating jet dumps bedload and some suspended load. Mouth bars are not decorative. They are architectural. They build upward until they interfere with the flow enough to force a decision.

To see why, imagine the channel as a pipe that is gradually being plugged. The river cannot simply accept the plug and rise like water behind a dam, because the river mouth is not a sealed container. Water will find a way around. As a mouth bar grows, it raises the bed elevation in the center of the channel just downstream of where the flow exits confinement. The deepest, fastest part of the flow, which typically runs down the channel center, is forced to split and move around the bar's flanks. Two jets form instead of one. Each jet scours a path around the bar, deepening it and encouraging more flow to take that route. In other words, deposition creates the obstacle, and the obstacle creates the bifurcation.

Bifurcation is a technical word for a split into two branches. In rivers, splits are surprisingly delicate. A bifurcation can be stable for years, or it can quickly become lopsided, with one branch stealing most of the discharge and the other silting up. The outcome depends on a tangle of feedbacks: how the split is shaped, how sediment is distributed between branches, how easily each branch can erode its bed, how tides or waves push back, and how floods periodically reset the geometry. But whatever the long-term fate of a particular split, the basic mechanism is robust: a river entering standing water tends to build mouth bars; mouth bars tend to force splitting; splitting creates multiple outlets; multiple outlets spread sediment over a broader front; spreading sediment builds land in a widening arc.

One can say, loosely, that a delta is what happens when a river tries to keep going after it reaches the sea.

The word *delta* itself carries a historical anecdote, and it is worth pausing over because it hints at how long humans have watched this unraveling. Herodotus, writing in the fifth century BCE, described the Nile's sediment-built land projecting into the Mediterranean as resembling the Greek letter Δ. Whether he coined the term or popularized it, the comparison stuck. It also reveals something subtle: even in antiquity, the delta was recognized as a shape made by deposition, not merely a swampy accident at the coast. The Nile, after all, was not simply arriving at the sea; it was constructing the agricultural platform that made Egyptian civilization possible, renewing fields with silt when the annual flood was allowed to spread across the floodplain.

That older relationship between river and delta was intimate: water overtopped banks, left sediment behind, and built land broadly. Modern deltas often live under a different bargain, in which the river is asked to behave like an obedient canal. Yet even under levees, the physics at the mouth still insists on its own logic.

A vivid example comes from the lower Mississippi, where the river's desire to drop sediment at its mouth collides with a human desire for reliable navigation. In the nineteenth century, ships struggled to cross the shallow, shifting bars at the river's passes into the Gulf of Mexico. The bar was not a single fixed obstacle but an expression of the mouth-bar process: sediment accumulating where the river jet decelerated. James Buchanan Eads, an engineer with a taste for audacious solutions, proposed in the 1870s to build jetties that would confine and focus the flow at one pass. The idea was simple: if the problem is that the river spreads out and slows, forcing it to stay narrow should keep it fast enough to scour a deeper channel through the bar. The jetties worked, dramatically. They did not repeal the river's tendency to deposit sediment at the mouth, but they changed where and how that deposition occurred by altering the jet. They were, in effect, an attempt to impose tributary-style confinement on a place where the river naturally wants to unravel into distributaries.

The jetties did something else, too, which is often the hidden story of delta engineering: by privileging one outlet and deepening it, they nudged the distributary system toward an imbalance. Any time discharge is concentrated, sediment is concentrated, and the land-building work is concentrated. That can be a gift to one shipping channel and a loss to the surrounding wetlands that would otherwise have received sediment. The river's splitting is not merely a geometric curiosity; it is how sediment is rationed across the delta plain.

The most striking thing about this reversal from tributaries to distributaries is that it is driven by a failure of transport. A tributary system is what happens when water is energetic enough, relative to slope and channel size, to keep moving downstream along one main route. A distributary system is what happens when that route self-sabotages by building up sediment at its downstream end. The river reaches its base level, the lowest elevation it can effectively erode to (often sea level), and the job changes. Upstream, the river cuts down into rock and soil and exports the debris. At the coast, it becomes a builder, stacking that debris into bars, levees, and islands.

This building is not gentle. The word *deposition* can sound like dust settling on furniture. Delta deposition is more like a construction site that never closes. During floods, the river delivers enormous pulses of sediment; the water can be thick with sand, silt, and clay, moving in a chaotic three-dimensional swirl. The channel bed can rise measurably during a single event. Mouth bars can grow in steps, with each flood adding a layer. Between floods, waves, tides, and currents can rework the deposits, trimming one edge and extending another. The shoreline is not a line but a zone of negotiation.

If you stand on a delta and look closely, you can see how the distributary network tries to solve a practical problem: how to get water and sediment across an increasingly flat surface without paying too much energy to do it.

On a steep slope, gravity is generous. A small drop in elevation provides the energy to keep flow moving, to maintain turbulence, to push grains forward. On a flat delta plain, gravity is stingy. The same discharge now has far less potential energy to spend per kilometer. The channel responds by becoming sinuous, by building natural levees along its banks, and by perching itself slightly above the surrounding wetlands. That perched state is common on deltas because channels are efficient at delivering sediment to their own margins; over time, the banks grow higher than the interior basins. The perched channel is both a triumph and a trap. It means the river has built its own elevated roadway, but it also means that when the roadway clogs or becomes inefficient, the river can more easily abandon it.

Even before abandonment, the distributary network is a way of spreading risk. A single channel forced to carry all the water to a single mouth is vulnerable to mouth-bar plugging and to the small accidents of sedimentation that can tip the balance between erosion and deposition. Multiple channels offer alternative routes. If one outlet shoals, another can take more flow. If one mouth bar blocks, another pass may deepen. A distributary network is, in that sense, a redundancy system built from sand.

This redundancy has a cost: every split dilutes the flow. After the first bifurcation, each branch carries only part of the discharge. Lower discharge tends to mean lower velocity and a greater tendency to deposit sediment in-channel. That can make distributaries shallow and temperamental. It is why deltas often require constant dredging for navigation and why river mouths can migrate or fragment over short time scales. The same mechanism that builds land also makes channels unreliable.

The unraveling is therefore not a single event at the shoreline. It is a cascade. Mouth-bar deposition forces a split; the split reduces the power of each branch; reduced power encourages more deposition; deposition encourages more splitting. The geometry becomes fractal-like, with similar patterns repeating at different scales: a major distributary divides into smaller distributaries, which divide again into narrow channels threading through marsh and mangrove. From above, it can look like a nervous system growing into the sea.

Yet the distributary network is not only a pattern; it is an accounting system for sediment. Each grain has a fate: delivered to one lobe of coastline or another, stored in a mouth bar, trapped in a levee, buried in a marsh, or carried beyond the delta into deeper water. The shape of the network determines those fates by partitioning the flow. A small change near the apex of the delta, where the main channel first begins to branch, can alter the nourishment of vast areas downstream. People living on deltas encounter this as a moral problem as much as a hydraulic one: who gets land-building sediment, and who gets saltwater?

The river itself has no opinion. It follows gradients of water surface elevation, and it reacts to friction and sediment in the channel. But the inversion from tributary to distributary makes the downstream end of the river unusually sensitive, because the system is now operating close to the threshold between transport and deposition. A little less slope, a little less discharge, a little more sediment, a little more vegetation resistance, a little more pushback from tides, and the channel can tip from scouring to filling. The delta is where small changes can have large geographic consequences.

This is why deltas feel alive even when they are quiet. A forested valley river can look stable for decades because its main work is hidden: slow erosion on hillslopes, slow meander migration on floodplains. A delta advertises its work. It is building out into open water, and the construction is legible: new islands, new sandbars, a mouth that shifts after a storm season, a channel that becomes too shallow to fish, a bay that turns brackish where it used to be fresh.

And it begins, again and again, with that moment when the river stops behaving like a collector and starts behaving like a divider: the jet spreading into still water, the loss of speed, the mouth bar rising, the flow splitting around an obstacle it made itself. What looks like chaos at the shoreline is often order of a different kind: not the order of convergence toward a single trunk, but the order of distribution, the river trying to lay down its load as widely as it can before the sea takes it away.

7.2 The Pendulum of the Plains

A delta does not only branch outward; it also *changes its mind.* The distributary maze can give the impression of a settled pattern, a botanical flourish fixed in mud. In reality, many deltas behave like a slow pendulum: the main river channel occupies one corridor for centuries, builds a lobe of new land, then abruptly abandons that investment and swings to a different corridor, starting the work again somewhere else.

That swing is called an *avulsion*. The word is blunt for good reason. An avulsion is not the gentle side-to-side migration of a meander; it is a sudden rerouting of the main flow into a new channel, often during a flood. The old channel may remain as a waterway, but it is demoted. The new channel becomes the primary artery, carrying most of the discharge and most of the sediment. If you live on the wrong side of the change, the landscape can feel as if it has been rearranged overnight: water levels that were predictable become strange; wetlands freshen or turn salty; a community that depended on a navigable channel finds itself on a backwater.

The unsettling part is that avulsion is not a rare accident. It is one of the central ways deltas build. A delta lobe protrudes into a lake or sea as sediment is delivered to a particular set of mouths. Over time, the protruding land changes the local slope of the river. The river, which is ultimately a machine for spending gravitational potential energy as efficiently as possible, finds that the route it has been using becomes less rewarding. It has lengthened. Its gradient has flattened. Friction has become more expensive per metre of progress. The river begins to search for a cheaper path.

The search is not conscious, of course. It is written into physics. Water flows downhill, and in doing so it continuously compares alternatives: any small low spot, any weak point in a natural levee, any shortcut across the delta plain that offers a slightly steeper water-surface slope. A delta is flat enough that small differences matter. A drop of a few centimetres over a kilometre can decide where the next flood goes.

The key term is *slope advantage*. Suppose a river has built a long, gently sloping channel out to a distant mouth. Somewhere upstream, there may exist a shorter route to the sea: perhaps a low swale leading to a nearby bay, or an older abandoned channel that still sits like a groove in the landscape. The shorter route has a larger slope for the same overall fall in elevation. Water prefers it because it can move the same discharge while spending less energy on friction. Once a trickle begins to use that shortcut during a flood, it can enlarge the route by erosion, and the river can tip rapidly from mostly-old to mostly-new.

Deltas make this easier by building their own traps. As sediment spills out of the channel during floods, it piles up along the banks, forming *natural levees*. These are not engineered embankments but ridges of sand and silt that accumulate because floodwater slows abruptly as it leaves the confined channel. Natural levees are higher than the surrounding marsh or backswamp, so the channel becomes perched on a low ridge. This is a clever way for the river to keep its own pathway open, but it also sets the stage for a dramatic failure: when a perched channel overtops its levee, the spillway can carry water onto the lower delta plain with a slope advantage. If the spillway finds a route to deeper water, it may not stop at being a spillway.

One can think of a delta as an elaborate compromise between two kinds of work. There is the work of *conveyance*: moving water and sediment to the coast. And there is the work of *construction*: distributing sediment across the delta plain to build land. A single stable channel excels at conveyance but does construction in a narrow band. Avulsions periodically reset the arrangement, allowing construction to occur in new places, feeding fresh sediment to wetlands that would otherwise starve.

The Mississippi River offers a famous, sometimes frightening illustration. South of the river's last bedrock bluffs, the land flattens and the river enters a long alluvial plain. For hundreds of kilometres it flows between self-built banks of silt and sand, above neighbouring fields and swamps. Along that reach, a large shortcut exists: the Atchafalaya River, which leads to the Gulf of Mexico by a shorter, steeper route. By the mid-twentieth century, engineers and local residents recognized that the Mississippi was increasingly tempted to switch course. The Atchafalaya was capturing more flow, deepening and widening as it did so. In the language of slope advantage, the Atchafalaya route offered the Mississippi a better bargain.

A full avulsion into the Atchafalaya would have created a new main river mouth hundreds of kilometres west of the present one, leaving the existing lower Mississippi as a diminishing distributary. That is not an abstract geomorphic curiosity. The lower Mississippi corridor supports major navigation, water supply, and communities; New Orleans sits within that precarious arrangement. In response, the United States built the Old River Control Structure in the 1960s to regulate the division of flow, deliberately holding the river in a partially restrained state. It is a remarkable admission of how powerful avulsion can be: a river that seems to obey levees and maps is, at the scale of decades, a system that must be actively prevented from taking the route it finds energetically preferable.

That modern intervention echoes an older truth about the Mississippi Delta itself: it is not a single delta, but a sequence of lobes, each one built when the main channel occupied a different corridor. The delta has repeatedly swung its sediment delivery east and west across the Louisiana coast, creating overlapping landforms whose ages can be read in subtle ridges, abandoned bayous, and the distribution of wetlands. The history is not a neat timeline, because channels can share flow and switch gradually before they switch decisively, but the pattern is clear enough to draw: lobe after lobe, like fingerprints layered over one another.

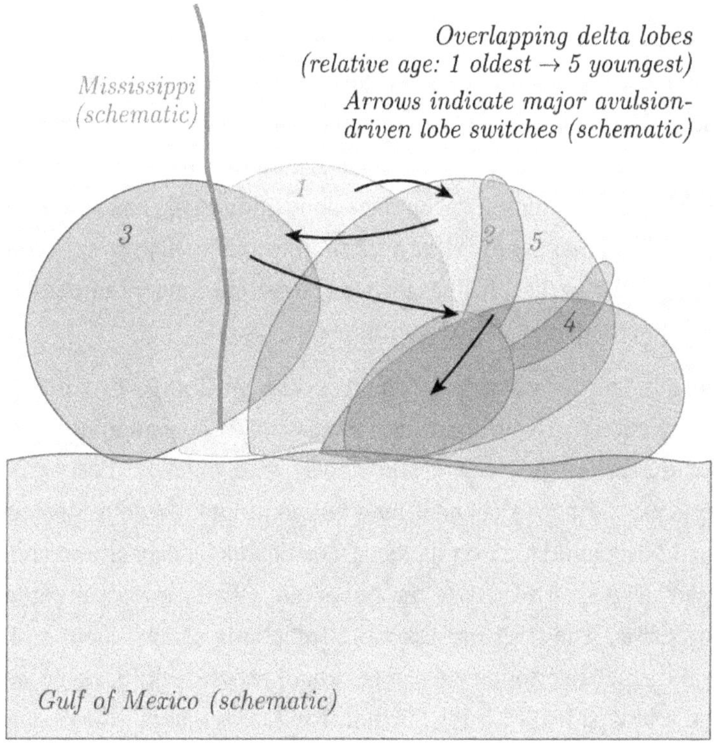

The drawing is deliberately schematic. The real Mississippi Delta is too intricate, and the names geologists assign to its lobes come with uncertainties and debates about exact timing. But the essential point survives simplification: the coastline of Louisiana is not the product of one steady river mouth. It is a palimpsest of several.

Avulsion is sometimes described as a catastrophe, and it often is for humans. But for the river, it can be an optimization. The old path becomes inefficient in a very specific way: it becomes too long for the slope it can command.

A simple way to feel this is to picture two ramps leading from the same height down to sea level. One is long and shallow; the other is shorter and steeper. Rolling a ball down the steeper ramp takes less time because gravity has a stronger component along the slope. Water behaves similarly, except that it drags against its boundaries and carries sediment that can clog the route. A delta lobe extends the long ramp farther and farther seaward, flattening the gradient near the mouth. That flattening encourages deposition in the channel, raising the bed and making the river even more perched. The river becomes like a delivery truck that has insisted on using a longer route and now finds the road slowly filling with its own cargo.

The moment of avulsion often arrives during a flood, when discharge is high enough to overtop natural levees and explore alternative paths. But the preconditions are laid down more quietly. For years or decades, the active channel may be aggrading, meaning that sediment is accumulating on the bed. The channel's capacity to convey water decreases relative to the floodplain. Water levels during high flows rise. The river presses more frequently against the top of its banks. Each overtopping event is a small experiment: does the water find a low pathway across the plain, and does that pathway connect to a lower outlet?

When the experiment succeeds, feedback accelerates everything. Once water begins to take the shortcut, it can erode it, deepening and widening it. Deeper channels carry more flow; more flow erodes faster. The old channel, robbed of discharge, loses its ability to keep sediment moving and begins to fill. This is the harsh beauty of avulsion: the new channel is carved not only by water, but by the *redistribution* of erosive power across the network.

Behind this drama is a quiet length scale that helps explain why avulsions tend to occur where they do. The lower reaches of a river approaching the sea are influenced by the water level at the receiving basin. When sea level (or lake level) is high, it acts like a downstream control, backing water up into the river. The reach where the downstream water level significantly affects the river's flow is called the *backwater zone*. In that zone, the river's depth and velocity respond not only to slope and discharge but also to the pushback from the sea.

Backwater matters because it changes where sediment is likely to accumulate. When a river enters the backwater zone, it often slows slightly and deepens. That can encourage deposition on the bed, especially during falling flood stages when the discharge drops but the channel remains deep. Over time, this can create a persistent tendency for the channel bed to build up in a particular band of the delta plain. Michael Lamb and Phairot Chatanantavet (2012) argued that avulsions commonly cluster around the backwater length: the distance upstream over which the sea's influence is felt. The idea is not that the backwater zone *causes* avulsion directly, but that it establishes a recurring location where aggradation makes the channel especially prone to spilling and switching.

This is one of those results that feels satisfying because it turns a messy natural history into a physical expectation. Deltas can look idiosyncratic, but they are not entirely free to do anything. Their restlessness has a geometry. The river is not wandering at random; it is repeatedly tripping over the same kind of problem in the same kind of place: the region where it is trying to deliver sediment efficiently while also being slowed by the sea.

The Mississippi's tendency to swing, and the Atchafalaya's temptation, also show why avulsion is inseparable from politics. The river's slope advantage is a natural calculation, but the consequences are distributed unevenly. A community on an aging lobe may depend on fresh sediment to maintain wetlands that blunt storm surge; when the river's main flow swings away, saltwater can intrude and marshes can thin. Meanwhile, a community near the new lobe may see land building, new channels, new fisheries, and new hazards. Avulsion is a reallocation of land-making capacity.

Engineers sometimes try to replace the pendulum with a lever: hold the main channel in place and distribute sediment deliberately through controlled diversions. That strategy is often discussed for coastal Louisiana, where the loss of wetlands has made the stakes painfully clear. Chris Paola and Robert Twilley (2011) framed this kind of restoration as an attempt to work with delta physics rather than against it: redirect water and sediment into adjacent basins so that the delta can build land where it is needed. The difficulty is that a diversion is not merely a pipe opening. It is a new branch in a network, and the network responds. Sediment does not distribute itself politely; it follows the altered hydraulics, builds bars, changes local slopes, and can even set up conditions for unintended channel shifts.

One of the most useful mental models here is to think of a delta as a network that routes fluxes: water flux, sediment flux, and even salinity flux. A change near the top of the network can ripple outward. Alejandro Tejedor and collaborators (2015) showed how representing deltas as directed graphs can clarify this: certain links and junctions disproportionately control how much flow reaches particular outlets. In everyday terms, some seemingly minor channels are load-bearing. If they fill with sediment or are blocked, large areas downstream can be starved. If they are enlarged, they can draw flow away from other parts of the coast. The pendulum is not a single swing from left to right; it is a continual negotiation among competing routes, with avulsion as the decisive punctuation mark.

For people living on deltas, the most disorienting aspect is timing. The slow preparation can lull you. A channel bed rises millimetre by millimetre; a natural levee grows imperceptibly; a backswamp subsides. Then a flood arrives, and water finds a new line of weakness. The switch can happen fast enough to be witnessed, mapped, argued over, and legislated about within a single lifetime.

Mark Twain famously joked that the Mississippi is always changing its mind, as if it had a personality. The humour lands because it captures the lived experience of a river that refuses to stay put. But beneath the joke is a stricter logic. The river is not fickle; it is economical. It builds land by delivering sediment, and in doing so it flattens its own slope, making its delivery route less efficient. Avulsion is the river's way of escaping the trap it has constructed.

A delta, then, is not only a place where a river unravels into many channels. It is also a place where the river periodically gathers itself and chooses, with startling decisiveness, a different way to be many. The swing leaves behind abandoned corridors that become bayous, lakes, and marshes; it opens new corridors that rapidly become the focus of deposition and land building. Over centuries the coastline becomes a record of these decisions: overlapping lobes, each one a frozen moment of advantage, each one slowly softened by tides, storms, and subsidence, while the river continues to seek the steepest bargain it can find.

7.3 The Geometry of Conflict

Stand at the mouth of a great river and the scene can feel like an argument made visible. The river arrives brown and pushy, full of suspended mineral dust and the occasional uprooted tree. The sea answers in a different language: a pulse that rises and falls, and a sideways shove that never quite stops. To a first approximation, a delta is the shape that remains after years of this brawling.

It is tempting to treat the shoreline as a line the river draws. Rivers do draw, in a way: they deliver sediment and extend land outward, and on some coasts the river's hand is unmistakable. Yet the river is rarely alone. Waves can redistribute sediment along the shore like a conveyor belt. Tides can pump water in and out of channels, deepening some and silting others, turning a neat fan into an estuary-like lattice. Even the weight of the delta itself can pull the ground downward as soft sediments compact. The coastline becomes a record not only of supply, but of theft: every mechanism that removes, rearranges, or submerges what the river tries to build.

The conflict has a geometry because each force prefers a different kind of shoreline.

A river, left to itself, tends to build outward in the direction of its momentum. That outward push can create protruding lobes, narrow fingers, and multiple mouths. The river is a delivery truck that wants to park as close as possible to the edge and unload. Waves, by contrast, tend to straighten coastlines. They are suspicious of bumps. A bulge into the sea becomes a target for oblique wave attack: waves refract (bend) as they approach shallow water, and the resulting currents often move sand from protruding headlands toward sheltered bays. Tides have their own preference. Strong tides carve wide inlets, deepen channels, and encourage sediment to settle in long, linear bars aligned with tidal flow. A tide-dominated delta often looks less like a bird's-foot and more like a set of funnels and sand ridges, with wide mouths that breathe with the daily cycle.

The geologist William E. Galloway, writing in the 1970s, popularized a simple but durable way of organizing these shapes: deltas can be described in terms of the relative dominance of river processes, wave processes, and tidal processes. It is not a law of nature, and real coasts sit on a continuum, but the framework works the way a good dinner-party analogy works: it gives you something to point at.

Consider three familiar impressions. A river-dominated delta can form a jagged, branching protrusion because the river's sediment delivery outpaces the sea's ability to rearrange it. The classic example is the modern Mississippi bird's-foot pattern, where narrow distributaries extend into the Gulf like outstretched fingers. A wave-dominated delta tends to form a smoother, arcuate (bow-shaped) shoreline. Sand is moved along the coast and spread out, so the river mouth becomes one contributor to a broader beach-ridge system. The Nile, especially in its more recent, wave-reworked form along the Mediterranean, is often cited as a wave-shaped counterpoint. A tide-dominated delta, such as parts of the Ganges—Brahmaputra—Meghna system, is threaded by wide channels and tidal creeks, with large islands and sand bars arranged by the in-and-out flow.

Those are impressions, not verdicts. The same delta can be river-dominated during a period of high sediment supply and relatively weak waves, and more wave-shaped at other times. A stormy decade can rearrange a coastline with a vigor that makes the preceding calm years look irrelevant. Still, the end-members clarify why deltas differ so dramatically in outline even when the rivers feeding them are comparable in size. It is not only how much sediment arrives; it is how aggressively the sea edits.

The editing begins with a deceptively simple act: a wave breaking at an angle.

When waves approach a shore obliquely, the swash and backwash of water on the beach tends to move sand along the shoreline, a process called *longshore transport*. One day's worth may be modest, but over years it becomes a relentless sideways drift. If a river delivers sand at one mouth, that sand can be carried down-coast, nourishing beaches far from the river itself. This is one reason deltas can influence regions well beyond their immediate footprint. The same mechanism can also starve areas up-coast if sediment supply is interrupted, as when dams trap sediment upstream or jetties block alongshore movement.

Longshore transport also explains why waves dislike bumps. A protruding lobe sticks out into deeper water where waves have more energy. Waves can attack its flanks and move sand away. Over time, protrusions are planed down and spread out. A wave-dominated delta becomes smoother and often more stable in outline, even though the sand within it is constantly in motion.

But waves can also do something more surprising: they can amplify bumps instead of smoothing them. This is not intuitive if you think of waves as sandpaper. Yet under certain wave climates—particularly when waves commonly arrive at high angles relative to the shoreline—the alongshore transport can become *antidiffusive*. Instead of smoothing irregularities, it can sharpen them, encouraging the shoreline to organize into large-scale features like capes and flying spits. Andrew Ashton and Brad Murray (2006) showed in modelling work how oblique wave incidence can lead to self-organized shoreline shapes, with sand concentrating into protrusions separated by embayments. The result is a kind of pattern formation: the coast develops a rhythm of capes and bays even without any obvious pre-existing template.

This matters for deltas because the delta front is often a source of sand to the littoral system. If the wave climate tends to produce instabilities, the sand delivered by a river might be sculpted into a series of promontories and spits rather than an evenly smoothed arc. The river may be doing the supplying, but the sea may be doing the composing.

Tides add a different kind of complication. A tide is not merely a vertical rise and fall. It is a horizontal exchange of water that can be enormous in volume, particularly in estuaries and wide inlets. That exchange generates strong currents that can carry sediment, scour channels, and build elongated bars. The key measure is tidal range and the strength of the associated currents; where those are large, the delta's distributary channels are not simply outlets but two-way conduits.

In a tide-influenced delta, the concept of a river mouth as a one-way jet becomes less reliable. During ebb tide, river and tide flow seaward together, accelerating currents and potentially scouring. During flood tide, seawater pushes landward, raising water levels in channels and sometimes reversing flow near the mouth. This can trap fine sediment, promote deposition in certain reaches, and alter where bifurcations remain open or close. The network becomes a breathing structure.

Tidal dynamics can therefore shape not only the coastline but also the distributary architecture inland. A bifurcation that would be unstable under purely river flow may be stabilized by tidal pumping that keeps both branches flushed. Conversely, tidal backwater effects can raise water levels and encourage sedimentation in places that make channels more prone to plugging. The delta becomes an arena where the river's desire to deliver and the ocean's desire to exchange collide in the channels themselves.

The conflict is not merely between river and sea. It is also between the surface the river is trying to build and the ground's tendency to sink.

Fresh deltaic sediments are often watery, loosely packed mixtures of sand, silt, clay, and organic matter. Over time they compact under their own weight. Pore water is squeezed out; grains settle into tighter arrangements. This *subsidence* can be slow, but it is persistent. It means a delta must build not only outward but upward, continually adding material to keep its surface near sea level.

Subsidence can be accelerated by human activity, particularly the extraction of groundwater, oil, or gas, which reduces pore pressure and allows sediments to compact faster. It can also be exacerbated by engineering that prevents floodwaters from spreading across the delta plain. If the river is confined behind levees and its sediment is routed directly offshore, the wetlands and floodplains that need mineral input are starved. Even if the delta front is gaining land in some places, the interior can be losing elevation relative to sea level. The shape you see from above can conceal a deeper loss: land turning into water not because it has been eroded, but because it has sunk.

The most painful demonstration of this comes from coastal Louisiana. Large areas of wetland have converted to open water over the past century. Many factors contribute, and debates can become political, but the physical ingredients are clear: reduced sediment supply to wetlands, subsidence, and a rising global sea level. The river is still delivering sediment, but much of it is routed through maintained channels and off the continental shelf, while the delta plain behind the levees is deprived of regular replenishment. The conflict there is not only a contest of planform shape; it is a contest of vertical survival.

It is worth pausing on a historical vignette because it shows how old this conflict is, even if the details have changed. In the Netherlands, centuries of living on low, soft ground at the edge of the sea produced an intimate knowledge of subsidence and storm-driven flooding. Dutch engineers and communities built dikes and polders to keep water out and make land usable. That effort is often celebrated as triumph over the sea, and in many ways it is. Yet it also demonstrates a recurring bargain: once water is excluded and land is drained, organic soils can oxidize and compact, and the land surface can sink. The more thoroughly you dry and control a deltaic plain, the more you may increase its dependence on continued protection. Control can create fragility.

Back at the shoreline, the geometry of conflict can be read as a balance sheet.

On the credit side is sediment supply: the amount and grain size of material the river brings, and how efficiently the distributary network delivers it to the parts of the coast that need it. On the debit side are all the processes that remove sediment from where it would build land: wave-driven redistribution, tidal export, storm erosion, and the slow but relentless lowering of the land surface by subsidence and relative sea-level rise. The balance sheet can vary across a single delta. One lobe may be accumulating land while another erodes. A mouth may prograde (advance) while a nearby bay opens wider.

This unevenness is where network geometry becomes more than a pattern. A distributary system partitions water and sediment among outlets. If waves and tides remove sediment from one part of the coast faster than it is supplied, that segment retreats. If an outlet receives a disproportionate share of sediment, it can build seaward even in a relatively energetic marine environment. Alejandro Tejedor and colleagues' graph-based view of delta networks makes this intuitive: change the capacity of a few links, and the downstream nourishment changes dramatically. A delta is not a uniform machine; it is a routing problem with winners and losers.

Deltas also carry their conflict offshore, into plumes and bars that are visible from space. After a flood, a sediment plume can stretch tens of kilometres into the sea. Some of that sediment settles close by, feeding mouth bars and nearshore deposits that can later be reworked onto beaches. Some is carried along the coast. Some is carried into deeper water and effectively lost to land building on human timescales. Whether sediment is retained near the shore or exported depends on grain size, wave energy, shelf currents, and the stratification between fresh river water and saline sea water. A delta that looks like it is growing may still be leaking its finest material to the deep ocean, while the coarser sand circulates in the nearshore zone.

There is a final twist to the geometry: storms.

Storms are not simply bigger waves. They can temporarily rewrite the rules by raising water levels (storm surge), allowing waves to attack farther inland, and by generating strong currents that breach barriers and open new inlets. A barrier island chain in front of a delta can be broken and reformed; a tidal inlet can migrate; a bay can be opened to the sea. In storm-prone regions, the "average" wave climate can be less important than the handful of extreme events that do most of the rearranging. A delta's shape is therefore partly a memory of rare violence.

If deltas inspire both awe and anxiety, it is because they sit at the intersection of human time and geologic time in an unusually direct way. A mountain erodes on a timescale that can feel abstract. A delta can change enough within a human lifetime to be witnessed and mourned. A fishing ground becomes too shallow to reach. A village that was once inland becomes waterfront. Salt kills trees that used to stand in fresh marsh. At the same time, the processes responsible are not exotic. They are the everyday physics of water moving sediment, scaled up and forced into conflict by the meeting of river and sea.

The shoreline is the public face of that conflict, but the conflict's geometry extends inward, into the branching network and the perched channels, and downward, into the compaction of mud. The river pushes outward, waves sweep sideways, tides breathe in and out, and the ground itself slowly lowers under the weight of what has been built. The delta persists not by finding a permanent shape, but by constantly renegotiating what counts as land at the edge of water, grain by grain, tide by tide, storm by storm.

Chapter 8

The Engineered Web

We built civilizations by taming rivers, transforming them from dynamic, shifting arteries into static, predictable pipes. But imposing rigid geometry on a chaotic network introduces a dangerous paradox: the infrastructure designed to protect us often creates brittle systems where safety in one location guarantees catastrophe in another.

8.1 The Concrete Straitjacket

A river in flood is often described as "overflowing its banks," as if the banks were the edge of the river and everything beyond were an accident. In reality, a floodplain is part of the river's operating space: a broad, low-lying area that accepts water when the channel cannot, spreading a surge outward and slowing it down. That spreading is not a side effect. It is a form of communication. When the river connects laterally—sideways—to wetlands, backwaters, and low fields, it trades speed for area, and kinetic energy for friction. The floodplain is where the flood wave is negotiated into something less sharp.

Engineered rivers are frequently designed to silence that negotiation. Levees, floodwalls, riprap, concrete linings, and straightened channels aim to keep water in a narrow corridor and move it along quickly. The intention is understandable: protect lives and property where people have built close to water. But the physics is stubborn. Constraining a flow does not remove its energy; it often concentrates it. A river confined between smooth, steep banks is like a crowd pushed into a hallway: the same number of people pass a given point, but the motion becomes faster, more forceful, and less forgiving. If the hallway ends abruptly, the consequences are not gentle.

Two ideas help make this tangible. The first is discharge, the volume of water passing a point per unit time, often written as Q. The second is the continuity relationship $Q = Av$, where A is cross-sectional area of flowing water and v is average velocity. If engineers reduce the effective area available to convey a flood—by raising levees that prevent spill onto the floodplain, or by narrowing a channel—then, for the same Q, velocity must rise unless water level rises enough to increase A within the confined corridor. In practice, both often happen: the flow speeds up and the flood stage climbs. Higher velocity increases the river's ability to erode banks and bed, while higher stage increases pressure on levees and floodwalls. The system becomes more brittle just as it appears more controlled.

The floodplain also provides storage. Think of a flood wave as a moving bulge of water traveling downstream. In an unconfined valley, part of that bulge spills outward, temporarily parked on the floodplain. The river pays a small delay and a reduction in peak height in exchange for a wide, shallow spread. The bulge returns later, draining back as the main channel falls. In a leveed valley, the bulge cannot park; it must remain in motion. That means the wave stays taller and sharper. Hydrologists call the river's water level through time a hydrograph. Confinement tends to make hydrographs "flashier": a quicker rise, a higher peak, and often a more abrupt fall. Flashiness is not just an aesthetic of graphs. It is stress: on bridges, on banks, on levees, on downstream towns, and on every organism that depends on calmer margins to survive a disturbance.

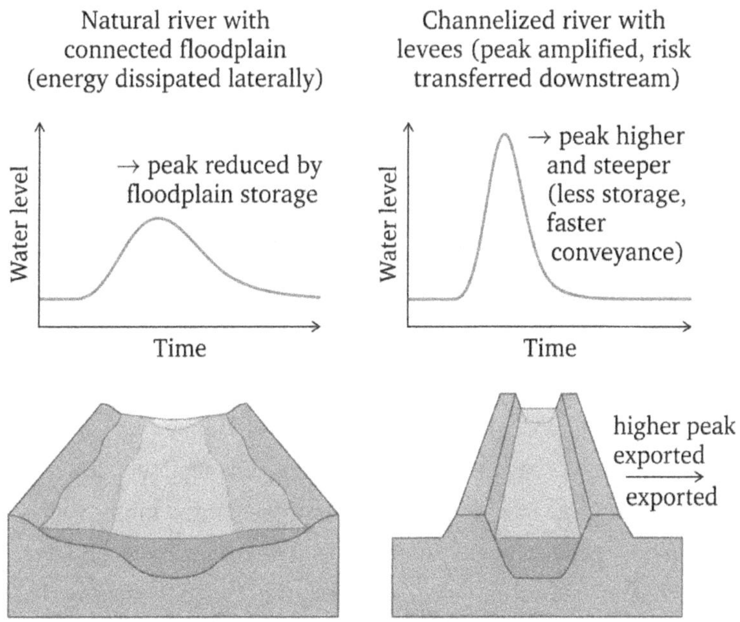

Cross-section: wide floodplain accepts water, increasing friction and storage. Blue: floodwater spread; Brown: banks/floodplain.

Cross-section: levees prevent lateral spill, raising stage and velocity in a narrow corridor.

There is a subtle moral hazard in the phrase "flood control." A river cannot be controlled in the sense that a thermostat controls a room; it can only be *redirected*. When water is kept off a floodplain at one reach, it must go somewhere else in the same time window. Often it goes downstream, faster. The community that sees the immediate benefit—a reduced frequency of nuisance floods, a cleared path for development—is not always the community that pays for the redistributed hazard.

The mechanics of this redistribution are easiest to see on rivers with long leveed corridors. Picture a flood moving down a valley. In a connected system, the leading edge of the flood spills onto low ground, where vegetation, microtopography, and shallow depths chew up momentum. In a leveed system, that edge is reflected inward. The water piles higher in the channel, increasing the water-surface slope, which is the river's downhill pressure gradient. Higher slope drives higher velocity, which in turn drives higher shear stress on the bed and banks. Shear stress is the river's ability to do work: to pick up sediment, to undercut revetments, to scour around bridge piers, to find the weakest point in a levee toe. What looks like protection can be an accelerator.

That acceleration often triggers a second process: incision. If a channel is straightened, smoothed, and forced to carry higher velocities more often, it tends to erode its bed. The channel cuts down, sometimes by meters over decades. At first glance this might sound like a win—a deeper channel, more capacity. Yet incision can lower the adjacent groundwater table, drying out wetlands and making riparian trees more vulnerable to heat and drought. It can also steepen banks, increasing the risk of collapse, and it can leave floodplains perched above the new channel, further reducing the frequency of beneficial overbank flows. The river becomes even more isolated from its valley. The disconnection becomes self-reinforcing.

A concrete channel adds its own twist. Natural channels are rough. Their beds are made of gravel, cobbles, logs, dunes of sand; their banks are indented, root-bound, and irregular. Roughness translates into friction. Engineers sometimes quantify this with a roughness coefficient (Manning's n), a number that, in effect, says how hard the river must "push" to maintain a given velocity. Concrete has low roughness compared to a vegetated floodplain or a cobble bed. When a channel is lined, the river can move the same discharge with a lower water level *in that reach*—the advertised benefit. But the water that arrives sooner and in a sharper pulse does not vanish. Downstream, where the river widens, bends, or meets a constriction, the energy has to be spent somewhere. If it cannot spill laterally because levees continue, it is spent vertically as stage, or destructively as scour and structural stress.

A historical example makes the tradeoffs less abstract. The Los Angeles River, now famous in films as a strip of concrete, was once a wandering, sandy river with broad washes and wetlands that braided and shifted across the coastal plain. In the early twentieth century, as Los Angeles expanded and paved over its permeable soils, flood peaks became more dangerous. After severe floods, federal and local authorities pursued large-scale flood control. In the mid-century build-out, long reaches of the river were channelized, lined with concrete or grouted stone, and paired with levees and flood-control infrastructure. The system did what it was designed to do: convey stormwater rapidly through a dense city and reduce flooding from frequent events.

Yet the design also created a new relationship between rain and risk. When hillsides and streets shed water quickly into storm drains, the river receives a sudden, synchronized delivery. The hydrograph spikes. In a natural setting, that spike might spread into wetlands and shallow basins, where it lingers and seeps down. In the engineered setting, the spike races. Residents do not experience the river as a gradual rise; they experience it as an abrupt hazard. The river becomes a place to be fenced off and warned against, not a living corridor whose margins can be inhabited safely at the right times. The city gains a sense of security inside the protected corridor, while the water's urgency is passed along the channel at high speed, making timing tighter and mistakes more consequential.

This is the social side of lateral disconnection: when the floodplain is taken out of the river's repertoire, people begin to treat the land behind levees as ordinary real estate. The most vivid expression of this is not a statistic but a familiar scene: warehouses and subdivisions pressed against a levee, children playing in backyards that sit below the river's flood stage, homeowners who have never seen water in the street and therefore assume water cannot arrive. A levee is a promise people want to believe, because believing it makes the world legible. The trouble is that a levee is better understood as a probability distribution built out of soils, maintenance budgets, vegetation management, burrowing animals, boat wakes, earthquake risk, and design assumptions about how big the next storm can be.

A river in its natural corridor fails gently. It spills, it spreads, it leaves silt and wrack lines and shallow ponds that drain over days. A leveed river fails sharply. Overtopping can erode the landward face rapidly; a breach can open a wide door; water that was riding high in a confined channel can pour into a bowl of development. The floodplain that once served as a rough, vegetated brake becomes, after development, a smooth basin filled with roads that guide flow and structures that trap debris. The same lateral connection that once dissipated energy becomes, under different land use, a pathway for concentrated damage. It is not the floodplain that is the enemy; it is the illusion that the floodplain can be occupied without cost because a wall exists nearby.

The phrase "risk transfer" can sound like jargon, but it captures a deep truth about networks: local improvements can create nonlocal problems. A levee that reduces inundation in one reach can raise stages in another by keeping more water in the main channel. Channel straightening that speeds flow through one city can deliver a higher, earlier peak to the next. Even within a single metropolitan area, one protected district can export hazard to another that lacks political power or tax base for comparable structures. Water follows gravity, but the choices that govern its pathways are made by institutions, property lines, and historical accidents of who built first.

There is also a quieter transfer: from common, frequent disturbances to rare, catastrophic ones. The natural floodplain expects to be wet from time to time. Its vegetation tolerates burial and saturation; its soils and microtopography are shaped by recurrent overbank flow. When levees eliminate smaller floods, that adaptive rhythm is broken. People grow less familiar with water. Emergency planning becomes abstract. Infrastructure is built under the assumption of dryness. Then, when an extreme event occurs—or when a levee fails—the landscape meets water without practice. The result is not just higher economic loss; it is deeper trauma, the kind that lingers in decisions about where to live, what to insure, and whether a community can recover.

A river's lateral communication is also a form of measurement. When water spreads onto a floodplain, it "reads" the valley: it encounters roughness, storage hollows, side channels, vegetation that changes with season, and groundwater that can accept some fraction of the flow. This reading feeds back into the shape of the channel over time. The river adjusts its width and depth, sorts its sediment, builds natural levees of silt along its banks, and leaves behind ridges and swales that guide the next flood. Engineering often replaces that adaptive geometry with fixed lines. The channel cross-section becomes a design template rather than a negotiated form. The river's energy, no longer allowed to spend itself laterally in small payments, accumulates as a debt that must be paid in fewer, larger transactions.

None of this means levees and channellization are always mistakes. In many places they have saved lives, protected critical infrastructure, and bought time for cities that grew before the costs were understood. The concrete straitjacket is, in part, a record of hard-earned fear: floods that drowned neighborhoods, swept away bridges, and left governments demanded to act. But fear can produce a narrow kind of intelligence. It sees the immediate threat and builds a wall at its edge. The river, meanwhile, remains a connected system, pushing against that wall with every storm, carrying the consequences downstream and into the future.

A more generous kind of intelligence begins with a simple question: where can the river be allowed to be a river? That question is less romantic than it sounds. It is about giving water somewhere to slow down without harming people: a park that floods, an agricultural field with compensation agreements, a restored wetland, a setback levee that creates room for a wider corridor. Each of these is a way of rebuilding lateral communication without pretending that cities can return to a pre-development landscape. The point is not to celebrate flooding; it is to recognize that energy must be dissipated, and that a floodplain is a tool as real as a spillway or a pump.

When a flood wave arrives, a connected river does something that looks almost like courtesy. It makes itself wide. It shares the load with grasses and trees, with shallow basins and old channels, with the slow exchange into groundwater. A confined river has no such manners. It stays narrow, rises higher, moves faster, and insists that someone else—often someone downstream, often someone with fewer defenses—should bear the force of water that was never meant to be kept in a single lane.

8.2 Arresting the Pulse

A dam looks, at first glance, like a simple interruption: a wall across a valley, a lake behind it, a controlled release below. The water still flows, so it is easy to assume the river still functions. Yet rivers are not made of water alone. They are made of *traffic*: water, yes, but also sand, silt, gravel, nutrients, wood, seeds, insects, fish, heat, and chemical traces from every hillside and street upstream. Water is the courier; sediment is the cargo that builds the channel and the floodplain. A dam changes the terms of delivery. It turns a moving river into a settling basin, and the first thing to fall out of suspension is often the river's ability to keep making land.

The phrase "sediment load" can sound like a nuisance, something that clogs reservoirs and muddles drinking water. From the river's perspective it is raw material. Sediment is how a river repairs itself after erosion, how it fills hollows, builds bars, nourishes deltas, and creates the shifting patchwork of habitats that many species use as a calendar. A flood that spills onto a floodplain does not just dissipate energy; it lays down a fine skin of silt that can fertilize fields and feed wetlands. Over centuries those thin skins become meters of soil. This is why so many old cities sit on river plains: not because people love the threat of flooding, but because rivers manufacture flat, fertile ground.

A dam arrests that manufacture by altering a basic balance: a river's capacity to carry sediment depends on its velocity and turbulence, while the amount of sediment available depends on what the basin supplies. Put a dam in the way and you create a sudden expansion from a narrow, fast channel into a broad, slow reservoir. When velocity drops, the river's carrying capacity drops. Sand settles first, then silt, then clay. Over time, the reservoir becomes a stack of the river's own building blocks, layered like a quiet geological archive. Downstream, the released water is often clearer and "hungrier"—a phrase engineers sometimes use, because clear water has spare capacity to pick up sediment. It erodes the bed and banks below the dam, scouring habitat, undermining bridges, and sometimes carving the channel deeper so that the floodplain is less frequently wetted. The dam traps the supply and may intensify the demand.

To grasp why this matters, it helps to picture a river as a conversation between *flow* and *form*. The form—the channel's width, depth, and pattern—evolves in response to how much water arrives, how quickly, and with what sediment. The flow, in turn, is shaped by that form: by roughness, bends, bars, vegetation, and floodplain storage. Dams intervene in that conversation at two points at once. They trap sediment, and they reshape the hydrograph, the time pattern of water levels. Those two interventions reinforce each other. A steady, regulated release tends to smooth the river's seasonal pulse; a sediment-starved release tends to simplify the channel that would otherwise have been rebuilt and rearranged by variable flows.

The smoothing of the hydrograph is often the main story told about dams: floods reduced, droughts buffered, electricity generated on demand. But the most profound ecological consequences often come from *timing* and *rate of change*. Many river organisms treat the rising limb of the hydrograph as a signal. A spring increase in flow can cue fish to migrate and spawn. A gradual fall after flood can leave side channels connected long enough for juveniles to grow. A natural river is not constant; it is rhythmically unreliable. That variability is not noise. As the ecologists N. LeRoy Poff and J. David Allan argued in 1997, the natural flow regime functions like a "master variable" that organizes habitats and life histories. When dams flatten the rhythm, the cues become faint or arrive at the wrong time.

Sediment is woven into that rhythm. Floods do not just bring more water; they bring a different *mixture* of particle sizes and organic material, and they do different work. A moderate flood might spread silt onto a floodplain; a larger flood might mobilize gravel, reshaping riffles and pools. A dam can reduce the frequency of the very flows that move coarse sediment, while still allowing flows strong enough to erode fine banks below the dam. The river becomes rearranged in a lopsided way: it loses building events and retains tearing events.

A concrete example makes the physics and the stakes visible. Consider the Nile. For thousands of years, the Nile's annual flood spread nutrient-rich silt across the valley and delta. Farmers timed planting to the water's retreat; the river's pulse was not just hydrology but national routine. In the twentieth century, a series of dams culminated in the Aswan High Dam, completed in the 1960s, which created Lake Nasser and largely eliminated the annual flood downstream. The benefits were real: more reliable irrigation, hydropower, and reduced flood damage. But the sediment that once replenished downstream fields and maintained the delta now settles behind the dam. Without that steady supply, the delta has become more vulnerable to erosion by waves and currents, and to subsidence—the slow sinking that occurs when delta sediments compact and are not replaced. When sea level rises, a sediment-starved delta has fewer ways to keep itself above water. The river has been asked to deliver water without land, and the bill comes due at the coast.

The Nile's story is not unique. The Colorado River once carried vast amounts of sediment to the Gulf of California, building and sustaining a delta of wetlands and riparian forests. With major dams and diversions, flows reaching the delta became dramatically reduced, and sediment delivery collapsed. What remains is a landscape that remembers being a river mouth—salt flats, abandoned channels, scattered wetlands sustained by return flows and occasional pulse releases—but no longer receives the regular materials that made it. When people speak of "restoring a delta" under such conditions, they are often really speaking about substituting an engineered pulse for a natural one, and about borrowing water and sediment from other needs.

One way to understand dams, then, is as devices that convert a moving river into a managed inventory. Water becomes a stock that can be held and released; electricity becomes a product; flood risk becomes a number; irrigation becomes a schedule. Sediment, unfortunately, becomes waste. Reservoir operators have reasons to treat it that way. Sedimentation reduces storage capacity, threatens intakes, and can damage turbines. Engineers dredge, flush, bypass, or simply accept the gradual loss of reservoir volume. Each of those choices is expensive, technically complex, or politically fraught. It is tempting to speak as if sediment were a maintenance problem inside the reservoir. But sediment is also the river's downstream pension plan: the deposits that keep floodplains fertile and keep deltas from retreating.

The interruption is not only physical but informational. A natural river carries a kind of message downstream: the imprint of storms, landslides, fires, and land use in the basin. After a wildfire, for example, sediment loads can spike; the river turns turbid, channel beds aggrade, and floodplains may be built. These are disturbances, sometimes dangerous ones, but they are also part of how a watershed resets itself. A dam muffles and delays such signals. The reservoir becomes a filter. Below the dam, the river's appearance can be deceptively stable: clear water, steady flow, predictable banks. Predictability is comforting, and it is one of the great selling points of regulation. Yet predictability can also mean that the downstream river has been cut off from the very variability that maintained its complexity.

Even when dams are operated with environmental intentions, the imitation is difficult. A release schedule can be tuned to mimic seasonal patterns, but it cannot easily reproduce the messy diversity of natural floods: the quick spikes and slow recessions, the unplanned sequences of events, the coincidence of sediment supply with peak flows. Water released from a reservoir can arrive cold or warm depending on intake depth, altering thermal regimes that fish have evolved around. It can arrive clear when a flood should be turbid, affecting feeding strategies and predator-prey dynamics. It can arrive at night or in winter for electricity demand rather than in spring for spawning. The river below becomes less a product of weather and more a product of human calendars.

The human story, too, is complicated. For many communities, dams brought electrification, industrial growth, and a sense of safety from floods. For others, they meant displacement, lost fisheries, and drowned homelands. The benefits and costs are rarely distributed evenly along the river. A city far upstream might enjoy hydropower and secure water supply; a village downstream might lose fertile silt, a seasonally flooded forest, or a fish run that once arrived like clockwork. The river's pulse is not just an ecological rhythm; it is a social contract, and dams rewrite it.

The pulse also has a geometry. Rivers are long, and distance matters. Sediment that would have traveled hundreds of kilometers may now be stored within a few. The downstream channel can respond in stages: immediate scour just below the dam, gradual bank erosion farther down, delta retreat at the far end of the system. Because these changes unfold over years to decades, they are easy to treat as unrelated problems: a bridge undermined here, a wetland shrinking there, a coastline eroding somewhere else. The dam sits upriver like a distant decision, while the river quietly rearranges itself around the missing sediment.

The most telling paradox is that dams often succeed at what they were built to do, while undermining the conditions that made that success meaningful. A flood-controlled river may protect a floodplain that has been converted to permanent infrastructure, but that same conversion can depend on a stable channel bed that sediment-starved water will not provide. A regulated river may supply irrigation water, but reduced overbank deposition can increase reliance on artificial fertilizers and the costs of maintaining soil structure. A delta may be protected by seawalls and dredged channels once sediment supply declines, but those protections require continuous investment and can never fully substitute for a living delta's ability to rebuild itself.

None of this requires imagining an impossible return to pre-dam rivers. It requires taking sediment seriously as part of river logic. Some systems experiment with sediment bypass tunnels, controlled flushing during high flows, or coordinated releases designed to mobilize and redistribute sediment downstream. These approaches acknowledge a basic principle: if a dam captures the river's building material, then downstream landscapes will eventually behave like landscapes deprived of construction supplies. They will erode, simplify, or sink, depending on local conditions. The question is not whether this will happen, but where and how fast, and who will be asked to adapt.

There is a final, unsettling lesson in the image of a reservoir slowly filling with silt. It reminds us that the river never stops working. Even behind concrete, gravity continues to pull particles downhill. The dam does not end the river; it forces the river to do its work in a different place, under different rules. Upstream, deposition accelerates, deltas form at the head of reservoirs, and tributaries back up into new wetlands that can be ecologically rich but socially contentious. Downstream, the clear water takes up the chisel, cutting into beds and banks that were once replenished. At the coast, where rivers once arrived heavy with sediment, shorelines become thinner and more negotiable.

A river's pulse is not merely the rise and fall of water level. It is the periodic delivery of matter that keeps landscapes from unraveling. When that delivery is interrupted, the downstream world does not simply become drier or safer. It becomes less able to hold its shape.

8.3 The Efficiency Trap

Engineers are paid to make systems behave. A river, by contrast, is a system that makes *itself.* It organizes channels, bars, and floodplains from whatever water and sediment arrive, and it does so without a blueprint. When people intervene, the instinct is to turn this unruly, shape-shifting thing into infrastructure: straighter, smoother, more predictable. The surprise is not that such interventions have side effects; the surprise is how often the side effects look like the very opposite of the intention. Flood control can make floods more catastrophic. Stabilizing flow can destabilize banks. Optimizing for reliability can produce sudden, system-wide failure.

The phrase "robust yet fragile" comes from network theory, the branch of applied mathematics that studies webs of connections: power grids, airline routes, supply chains, ecosystems, the internet. "Robust" means the system performs well under the disturbances it was designed to handle; "fragile" means it can fail dramatically under a disturbance outside that design envelope. Modern life depends on such systems. We rarely notice their subtle daily successes, but we remember their rare spectacular breakdowns.

It helps to start with a familiar analogy: an airline network. A hub-and-spoke system is efficient. It concentrates passengers through a few airports, reducing costs and increasing frequency on major routes. It is also fragile: fog at a hub can cascade into delays across a continent. A more redundant network with many direct connections can route around trouble, but it is expensive to build and operate. Rivers, especially engineered rivers, face a similar tradeoff between efficiency and redundancy, even though the currency is not passengers but water, sediment, and risk.

A natural river network contains redundancies of a particular kind. Water does not move only down the deepest channel. During high flows it spreads into secondary channels, backwaters, sloughs, floodplain swales, wetlands, and temporary ponds. Some of these pathways are intermittent, activated only when water levels cross a threshold. Others are slow, such as exchange with the subsurface. This patchwork looks messy on a map, but it is a form of insurance. If one pathway is blocked by a logjam or a landslide, others take up the load. If a flood arrives, the network enlarges itself laterally, distributing energy across more surface area and more frictional contact. In a literal sense, the river has multiple "routes" available.

Engineering often removes those routes. Levees keep water in the main corridor; channelization makes that corridor smoother; storm drains connect streets directly to channels; dams regulate flow into a narrow band of discharges; bank armoring prevents side channels from developing. The result is a network that looks, in graph-theory terms, more like a series of high-capacity links arranged in a line: efficient for conveyance, poor at detours. A system like that can be extraordinarily good at managing ordinary storms, and then abruptly overwhelmed by the combination of rare intensity and one local weakness.

The weakness is rarely mysterious in hindsight. It can be a low point on a levee crown, a poorly compacted section, an animal burrow, a tree uprooted in high water, a culvert that becomes clogged with debris, a pump that loses power, a gate that is not closed in time, a maintenance backlog that no budget quite covered. The distinctive feature of an efficiency-trap failure is not that the failure begins at a weak point—all structures have weak points—but that the consequences propagate far beyond it.

Network theory offers a simple vocabulary for this propagation. A network has *nodes* (places where things accumulate, branch, or are controlled) and *links* (pathways for flow). In a river system, nodes might be reservoirs, leveed reaches, confluences, floodways, tidal gates, stormwater outfalls, or simply constrictions like bridges. Links are river channels, engineered canals, bypass channels, and sometimes underground pipes. When the network is optimized for efficiency, it tends to concentrate flow through fewer links with higher capacity. That concentration makes the system easier to manage and cheaper to maintain on paper. It also increases *betweenness*—a network measure meaning that many paths depend on the same critical segment. High-betweenness links are the hubs of a linear system. They are where failure hurts most.

Rivers add a twist that power grids and airlines do not share: gravity enforces direction, and floods can create new links on the fly. When water overtops a levee, it does not merely "leave the network." It creates an unplanned route across land, often directed by roads, rail embankments, and the subtle slopes of developed terrain. That new route can return downstream in unexpected places or pond behind another embankment until it finds a weak point. The network is suddenly larger and less legible, but not in a comforting way. It is redundant in the wrong places and absent in the right ones.

The unsettling part is that many engineered river networks are designed as if the statistics of flooding are stable: as if the distribution of storm intensities, snowmelt timing, sea levels, and watershed land cover can be summarized by the past century of measurements. That assumption, called stationarity, once made practical sense. It allowed designers to speak of a "100-year flood" and build to a standard. But when climate shifts, when urbanization changes runoff, when forests burn more intensely, when sea level rises and backs water up into rivers, the distribution changes. A system designed to be robust to the "expected" is exposed to a growing set of "unexpected" events. Robustness becomes a kind of historical memory that the future does not respect.

The efficiency trap also has a social mechanism, not just a hydraulic one. Protection changes behavior. A levee that reduces frequent floods makes adjacent land more valuable and attractive for development. Over decades, this can create dense neighborhoods, industrial parks, and critical infrastructure behind a structure that was designed for a particular level of hazard. The residual risk—the chance that an event exceeds design, or that a structure fails in some way—does not disappear. What changes is the scale of loss when the residual risk is realized. An extreme flood striking a sparsely developed floodplain is a disaster; the same flood striking a fully built landscape behind levees is a catastrophe.

This is not simply a matter of individual choices. Flood insurance rules, zoning decisions, and political pressures all shape where development occurs. When maps draw a thick line around "protected" areas, the public reads that line as safety, even if engineers describe it as probabilistic. Over time, a protective structure can become a promise that no one remembers was conditional. The trap is sprung when a rare event tests that promise.

There is a painful American example that illustrates both the network dynamics and the social dynamics: the flooding of New Orleans during Hurricane Katrina in 2005. The city sits in a bowl between the Mississippi River and Lake Pontchartrain, with levees, floodwalls, canals, pumps, and gates forming a vast engineered network intended to keep water out and move rainfall away. The system performed well under many storms. Then Katrina brought an extreme storm surge, high water levels, and stress on multiple parts of the network. Breaches and overtopping at several locations allowed water to pour into neighborhoods, where it pooled in low-lying areas. The failure was not a single point but a cascade: once water entered, the bowl geometry and the reliance on pumps made drainage slow and vulnerable to power and mechanical constraints. The tragedy was magnified by the concentration of people and infrastructure in low elevations that had been made habitable only through continuous, coordinated engineering.

The details of Katrina involve engineering, governance, and human vulnerability in ways too extensive for a single telling, but one lesson fits the logic of networks. A city protected by interconnected structures can appear safe right up until it is not. The very integration that makes the system efficient in daily operation can make it brittle under compound stress. When multiple links are pushed near capacity at once—surge at one boundary, rainfall inside the perimeter, high river stages limiting discharge, communication failures slowing response—small weaknesses stop being local.

Another less famous but equally instructive example lies in the long, leveed corridors of large rivers such as the Mississippi. Along such rivers, raising a levee in one place can raise flood stages elsewhere by keeping more water in the main channel. This is not a moral indictment of any one project; it is a network effect. The river and its floodplain form a coupled system. Reduce storage at one node and the wave's energy and volume must be accommodated by other nodes, often downstream. Recent work describing "flood teleconnections" puts a modern name to an old observation: actions taken to protect one community can change hazard far away, through the network's hydraulic connectivity. The teleconnection is not mystical; it is simply the consequence of moving water faster through fewer pathways.

Teleconnections create a particular kind of fragility: political fragility. When benefits and harms are separated by distance, the feedback becomes weak. The protected community sees the immediate gain. The downstream community may not be able to point to a single project and say, "That is what raised our flood." Meanwhile, responsibility is fragmented across jurisdictions. Rivers ignore county lines, but budgets and accountability do not. A network can therefore be engineered into a state where each local optimization is rational and the global outcome is unstable.

Network theory also highlights the role of *thresholds*. In many systems, performance is smooth until a threshold is crossed, after which behavior changes abruptly. Levees have a threshold: overtopping. Reservoirs have a threshold: spillway activation or emergency releases. Urban drainage has a threshold: inlet capacity and pipe surcharge. When flows approach these thresholds, the system can shift from orderly conveyance to chaotic inundation. The irony is that efficiency often pushes a system to operate closer to thresholds more of the time. A channel designed to convey quickly can tempt planners to use its apparent spare capacity by building closer to it. A dam managed tightly for water supply can leave less storage for flood buffering. A pump network sized for typical storms can be overwhelmed by a rare cloudburst intensified by warmer air. The system becomes *tuned*—and tuned systems can be exquisitely sensitive.

It is tempting to respond by making everything bigger: higher levees, larger spillways, deeper channels, more pumps. Sometimes that is necessary. But scaling up within the same logic can deepen the trap. Bigger infrastructure can enable more development, increasing exposure; it can export hazard more forcefully; it can create a larger, more complex machine with more potential failure modes. Efficiency, in other words, is not just a property of water moving through concrete. It is a policy and cultural preference: the preference for singular solutions that maximize throughput and minimize visible inconvenience.

The alternative is not to abandon engineering, but to diversify pathways and to widen the margin between ordinary operation and catastrophic thresholds. In network language, that means introducing redundancy and reducing dependence on a few high-betweenness links. In river language, it often means restoring or preserving space where water can go without destroying what people need: flood bypasses that are activated intentionally, setback levees that create room for rivers to spread, wetlands that store surges, urban designs that slow and infiltrate runoff rather than rushing it into pipes, operational rules that keep some reservoir capacity available for floods, sediment management that maintains channel stability instead of forcing clear-water scour downstream. These are not aesthetic preferences. They are ways of changing the network topology from a brittle line back toward a more forgiving web.

Such choices also change what "success" looks like. An efficiency-minded system measures success as keeping water within a narrow corridor and meeting targets with minimal variability. A resilience-minded system measures success as avoiding catastrophic loss even if that means accepting small, planned inundations. The emotional hurdle is that planned inundation feels like failure, while a higher levee feels like action. Yet the river does not interpret a levee as action; it interprets it as a boundary condition. The river will test that boundary condition repeatedly, with a persistence that outlasts electoral cycles and maintenance schedules.

The deepest reason engineered rivers can become robust yet fragile is that they trade the river's distributed intelligence for centralized control. A natural floodplain stores, slows, and spreads water through countless small interactions: stems and roots, depressions and ridges, side channels and porous soils. None of these elements is critical on its own. Together they make a system that fails gradually. Concrete and steel replace that distributed work with discrete components: a wall, a gate, a pump station, a spillway. Discrete components can be monitored, regulated, and optimized. They can also break, and when they do, there may be no equivalent, quiet backup.

When people stand on a levee during high water, what they often feel is not only fear but admiration: the sense that a thin ridge of earth and concrete is holding back a continental force. That admiration is not misplaced. The structure embodies skill, labor, and public will. The danger lies in mistaking that ridge for a final answer. A river network that has been made efficient can look stable right up to the moment it flips into a different regime, when water finds a new path and reminds everyone that the most powerful routes in a network are sometimes the ones no one meant to build.

Chapter 9

Letting the River Be

The era of total hydraulic control is ending, revealing that rigid concrete often fractures under the weight of dynamic chaos. We uncover how 'un-engineering' our landscapes-removing dams and reconnecting floodplains-allows the inherent intelligence of the water network to heal itself. Ultimately, we find that long-term safety lies not in opposing the flood, but in designing our civilization to dance with it.

9.1 The Art of Un-Engineering

A river is one of the few everyday things that is never quite the same twice. The water changes hour by hour, the sediment load rises and falls with storms, trees drop into the current, banks slump, bars grow, and channels split or reunite. That restlessness is not a flaw in the system. It is the system.

Yet much of modern river management has treated change as the enemy. The instinct is understandable: a stable line on a map is comforting. A straight channel is easy to survey. A fixed bank is easier to insure. If a river migrates toward a town, the town does not want a lesson in geomorphology (the science of how landscapes are shaped); it wants the river to stop.

Traditional hydraulic engineering, at its most confident, tries to freeze a living process into a designed geometry. The vocabulary gives it away: *training* works, *control* structures, *armoring*, *bank protection*. The channel is told where to go and how wide to be. Gravel is instructed to stay put. Wood is removed before it can complicate the flow. The river becomes a machine that should run smoothly and predictably.

The trouble is that rivers do not merely *occupy* channels. Rivers *make* channels. The channel you see is a temporary arrangement negotiated between flow, slope, sediment supply, and the resistance of bank material and vegetation. Change any one of those, and the river renegotiates. When we lock a channel into place, we have not eliminated energy; we have redirected it. Often the energy reappears as deeper scour at the toe of a hard bank, erosion at the end of a revetment, incision downstream of a drop structure, or deposition where the river finally finds room to slow down. A design that "holds" at one point can export instability to the next.

Un-engineering begins with a different ambition. Instead of imposing a shape, it tries to restore the river's ability to do work: to erode where erosion is appropriate, deposit where deposition is appropriate, and build complexity from the inside out. The goal is not a fixed channel; it is a self-renewing set of processes.

That idea can sound suspiciously like giving up. It is not. It is closer to a shift in what counts as control. A rigid structure offers control over *form* (the river looks the way we drew it), but often at the cost of fragility (it depends on constant maintenance and on flood behavior staying within the assumptions of the design). Process-based restoration aims for control over *trajectory*: not "the river will be exactly here," but "the river will have room to move, and when it moves it will create habitats and reduce catastrophic surprises."

A useful term here is *dynamic stability*. A river can be stable in the sense that it does not blow itself apart, even while its channels migrate, split, and rejoin. The stability is statistical rather than geometric. Over years to decades the river maintains a characteristic width, slope, and pattern suited to its setting, but the details are always in motion. Many of the most diverse river habitats are not features that can be poured into place like concrete; they are the byproducts of that motion. Side channels form when floods cut across bars. Pools deepen where flow converges around a logjam. Gravel beds refresh when sediment can actually travel through the system.

Consider a common piece of "hard" river control: the grade-control structure. A grade is simply the channel bed elevation, and "controlling" it means preventing the bed from eroding downward. In many regulated rivers, the bed wants to cut down (incise) because upstream dams trap sediment or because the channel has been straightened and steepened, increasing its power. A grade-control weir or drop structure pins the bed like a thumbtack. Upstream, it may succeed for a time; downstream, the river often compensates by scouring a plunge pool, undermining banks, and creating a hungry reach that erodes until it meets another thumbtack. You can, in effect, end up building an expensive staircase to stop a river from doing what the physics keeps insisting on.

Un-engineering asks a blunt question: what if the thumbtacks are part of the problem?

That question does not require romantic faith in wilderness. It requires following the energy. Water flowing downhill loses potential energy. It must spend that energy somehow: on friction with the bed and banks, on turbulence, on moving sediment, on bending around obstacles, on spilling onto floodplains. If we smooth the channel, straighten it, and confine it between hard edges, we reduce the ways energy can be dissipated locally. The energy does not vanish; it concentrates. We get higher velocities, sharper peaks, deeper scour, and failures that arrive suddenly rather than gradually.

One of the most striking modern demonstrations of un-engineering is large dam removal. Dams are often discussed in terms of electricity or water supply, but geomorphically they are powerful *sediment traps*. Gravel, sand, and silt that would naturally feed downstream bars and deltas instead accumulate in the reservoir. Downstream, the water released from the dam is relatively sediment-poor and can behave like a solvent: it has capacity to carry sediment but little available to carry, so it erodes the bed and banks to make up the difference. Over decades the downstream channel can become narrower, simpler, and more incised, with floodplains disconnected by higher banks and lower water tables.

When a large dam comes out, the first thing everyone worries about is the slug of stored sediment. The worry is justified. Rivers are not polite about rearrangement. They can run thick with silt, bury habitats, and transform a mouth or estuary. The point of un-engineering is not to pretend those disruptions do not occur. It is to understand them as part of the system rebooting itself, and to decide when the long-term benefits of restored connectivity outweigh the short-term costs.

The Elwha River on Washington's Olympic Peninsula became a landmark case. Two dams built in the early twentieth century blocked salmon runs and trapped vast amounts of sediment. Their removal, carried out in phases in the early 2010s, released sediment and wood that had been stored for nearly a century. The river did what rivers do when given back their continuity: it cut into reservoir deposits, carried material downstream, braided and reworked its channel, and began rebuilding bars, floodplains, and the nearshore delta at the coast.

What makes this story more than a slogan is the timescale. The "movie" of a river responding to liberation does not run for an afternoon. It runs for years, then decades. Early on, the water can be muddy and the channel chaotic, a period that is unnerving if you are used to thinking of a river as infrastructure. Later, the same sediment pulse that looked like damage begins to look like raw material. Gravel and large wood help form pools and cover. New bars become seedbeds for riparian forests. At the river mouth, sediment can reshape beaches and shallow-water habitats. None of this follows a single neat template, because a river is not a factory assembly line; it is a network responding to local slope, valley shape, bank materials, and the timing of floods.

A quieter but equally telling example can be found not in removing a dam, but in removing the small constraints that make a channel behave like a ditch. Many streams were straightened for agriculture or flood conveyance, pushed into trapezoidal cross-sections, and kept there by levees or spoil berms. The result is often a channel that is too efficient: it rushes water downstream, erodes its bed, and then sits disconnected from the surrounding land except during the most extreme floods. Habitat diversity collapses because the channel lacks the features that create variety in flow depth and velocity.

Un-engineering in such settings often starts with giving the stream *room* and *roughness*. Removing or setting back a levee is not simply subtracting a wall. It is reintroducing a lateral dimension that allows floods to spread, slow, and deposit sediment rather than cutting deeper. Similarly, removing bank armoring and allowing some erodible banks can supply sediment that the channel needs to build bars and bends. A stream that can access its floodplain can also recharge groundwater locally, supporting riparian vegetation that, in turn, increases bank strength in a living, adaptable way.

The counterintuitive part is that letting a river erode in one place can make it less dangerous overall. When a river is confined, it often compensates by incising, which can lower the water table and weaken riparian forests; it can also create a narrow, deep channel that carries floods quickly, raising peak levels downstream. When the river is allowed to migrate within a wider corridor, it spends energy on reshaping its own boundaries. That "wasted" energy is exactly what makes the system less prone to catastrophic failure. A river that is permitted to fidget is often a river that does not save up its force for one devastating strike.

This is the logic behind process-based restoration: the design is not a fixed endpoint but a set of boundary conditions. Instead of sculpting a meander with excavators and hoping it stays put, practitioners might remove berms that prevent overbank flow, reintroduce large wood that encourages channel splitting and pool formation, or fill an artificially deep trench to raise the bed toward the floodplain. The interventions can be modest in appearance but profound in consequence because they change feedback loops. Once the water can spill, it deposits; once it deposits, vegetation colonizes; once vegetation grows, it adds resistance, which changes where sediment settles during the next flood.

There is a practical craft to this. "Letting the river be" does not mean ignoring human safety or property. It means choosing what to constrain and what to release, with an eye to how the river distributes its work. One can confine the river tightly through a town and then deliberately provide a wider corridor upstream or downstream where the river can dissipate energy and store sediment. One can protect truly critical infrastructure while allowing agricultural fields to flood occasionally in exchange for lower peaks and healthier ecosystems. The river becomes something like a partner in risk management rather than an adversary to be pinned down everywhere at once.

A persistent mistake is to imagine that nature-based approaches are always gentle and predictable. A river set free is not a garden hose; it is a force field. It will test banks, find weak points, and exploit them. That is precisely why un-engineering requires humility and patience. In a rigidly engineered channel, success is often judged by whether the structure looks unchanged after a flood. In a process-based approach, success can look like change that stays within acceptable bounds: a bend migrating but not threatening a road because the corridor was wide enough; a newly formed bar that creates shallow habitat and slows flow; a side channel that activates during high water and then quiets.

Those acceptable bounds are not purely technical. They are social decisions about what land is for, who bears the risk, and what benefits matter. A community may accept a river that moves within a designated corridor if the alternative is a cycle of escalating levees and sudden failures. Another community may not. Un-engineering works best where the geography and the politics both allow room.

A useful mental picture is the difference between building a canal and cultivating a braid. A canal aims for uniformity: a single, clean path, minimal friction, maximum conveyance. A braided or meandering river aims for adaptability: multiple pathways, lots of friction, constant rearrangement. The canal treats sediment and wood as nuisances to be removed. The living river treats them as building blocks.

Year 0: channelized corridor

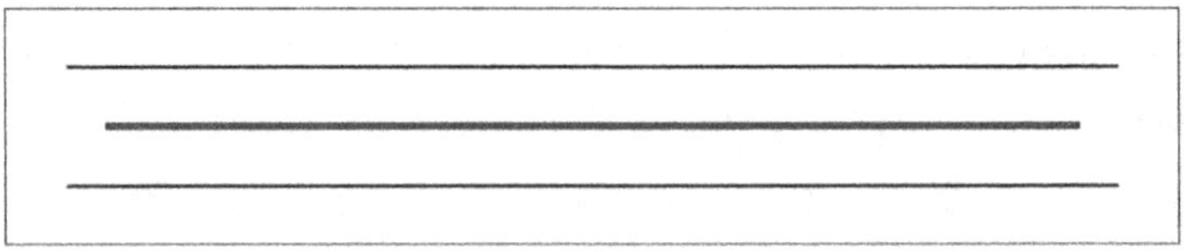

Year 5: constraints removed, channel begins to wander

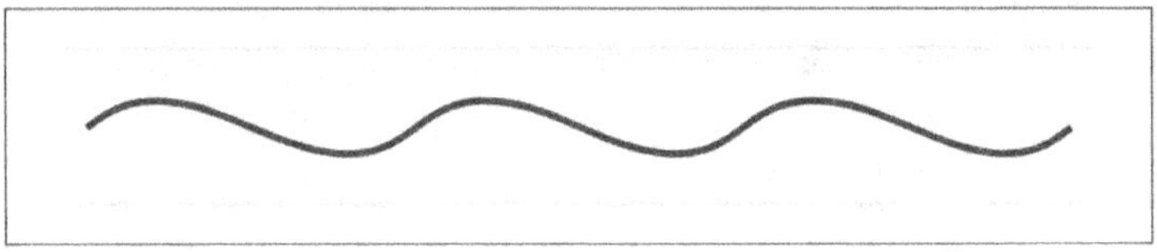

Year 50: complex planform with side channels and floodplain connectivity

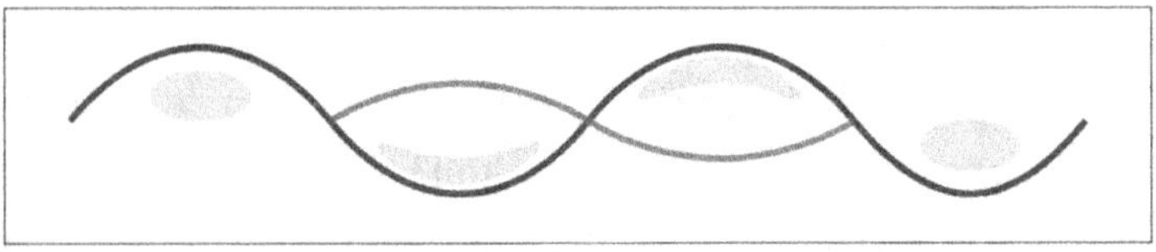

The diagram is deliberately simple, because the essential point is not the exact shape of a particular meander. It is the difference between a river forced into one geometry and a river allowed to search. In the first case, the "order" is imposed from outside. In the second, order emerges from the river's own adjustments.

One reason this approach can feel unsettling is that it changes what it means to finish a project. A concrete wall is done when the forms come off. A process-based restoration is "done" in the way a planted forest is done: the initial actions are complete, but the main work is carried out by time, floods, and growth. That does not make it vague. It makes it measurable in different ways: how often the river reaches its floodplain, how much sediment moves through, how diverse the flow depths and velocities become, how quickly habitats recover after disturbance, how maintenance needs change, and whether risk is reduced where people actually live.

The deepest appeal of un-engineering is not that it returns a river to some imagined pristine past. It is that it acknowledges a basic fact about networks shaped by gravity: energy will find a path. We can spend endless effort trying to dictate the path everywhere, or we can decide where we truly need rigidity and where we can afford-and even benefit from-flexibility. When a river is allowed to do some of its own design, complexity is not something we have to manufacture. It is something the flow builds, one flood at a time, out of the materials the watershed already provides.

9.2 Sponges, Not Walls

Floods become disasters less because water exists than because it arrives in the wrong *shape*. Shape sounds like a visual word, but here it means the curve of flow through time: how quickly a river rises, how high it crests, how long it stays elevated, and how fast it falls. A city can survive a day of moderately high water that spreads through parks and fields; it can be shattered by an hour of extreme water that overtops defenses and tears at foundations. The same total volume can be either inconvenience or catastrophe, depending on whether it is delivered as a spike or a swell.

That curve has a name: a *hydrograph*. On its simplest form, a hydrograph is a plot of discharge (the rate of flow, usually in cubic meters per second or cubic feet per second) against time. Discharge is not the amount of water a river *has*; it is the amount passing a point each second. The difference matters because floods are about rates. A bathtub does not overflow because it holds water; it overflows because water arrives faster than it can drain.

Engineered rivers often aim to evacuate floodwater quickly. Straightening, deepening, and lining channels lowers friction and increases conveyance. Locally, that can seem like success: the water level drops faster and the river appears obedient. But speed is a kind of debt. When the same flood volume is pushed downstream with less delay and less spread, it can raise the crest somewhere else, and it can do it with less warning. Smooth channels create smooth accounting for surveyors, not for hydrology.

Attenuation is the opposite of speed-as-a-solution. To *attenuate* a flood is to reduce its peak by stretching it out in time. The volume does not vanish. The hydrograph is reshaped. The peak comes down because the water is temporarily stored in places that are not the main channel, and because energy is dissipated by roughness: vegetation, small channels, shallow flow over uneven ground, and the chaotic turbulence of water spreading out rather than being forced into a single fast thread.

One can think of attenuation as the river doing what any crowded system does when it is given space. A hallway with a single narrow door produces a stampede. Open side doors into a courtyard and the pressure eases. In water, the "side doors" are floodplains, wetlands, bypass channels, backwaters, and low-lying fields that can hold water for hours to days and release it slowly. In many river valleys, those places used to be a regular part of the river's working life. Then they were walled off.

A wetland is often described as a sponge, and the metaphor works as long as it is not taken too literally. A household sponge soaks up water inside itself; a floodplain wetland "soaks" mainly by providing space for water to spread across the surface and into shallow storage. Some water does infiltrate into soil, and vegetation does transpire water back to the atmosphere, but those are not usually the decisive processes during a fast, large flood. The decisive process is temporary storage: water that would otherwise be stacked vertically in a confined channel is instead spread laterally over a wide area, lowering the height of the flood wave.

The physics is simple enough to explain at a dinner table, but it helps to name two ideas. The first is *continuity*: if a given volume must pass downstream, then lowering the peak means the passage must take longer. The second is *roughness*. Water moving over grass, shrubs, trees, and shallow depressions loses momentum through drag. A concrete channel is like a polished chute; a vegetated floodplain is like pushing through a crowd. The river does not need to be "stopped" to be made safer. It needs to be made slower in the right places.

The contrast becomes vivid when drawn. Two rivers can deliver the same amount of water past a point, but one does it as a narrow, dangerous peak and the other as a wider, lower crest. The area under each curve is the total volume that passed; the difference is how threatening the peak is.

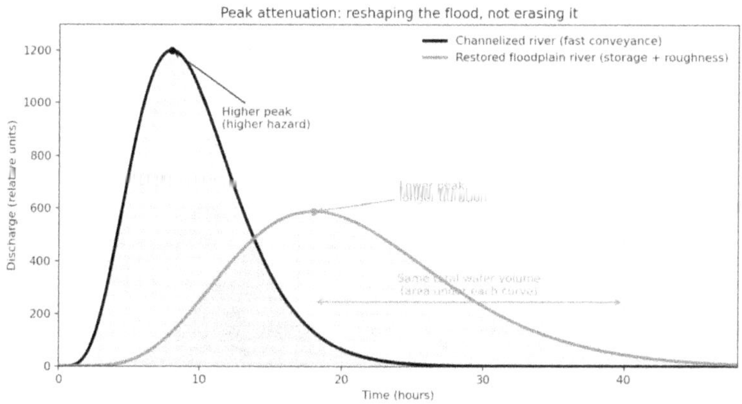

The plot hides a moral choice inside a simple geometry. The black curve is what happens when water is kept in one efficient pipe: fast rise, high crest, steep fall. The blue curve is what happens when the river has places to spill and slow: the crest lowers, the warning time increases, and the force against levees, bridges, and buildings can be reduced. A person standing by the river might still see "a flood" either way, but the city downstream feels the difference as either a manageable strain or a violent test.

Wetlands matter because they are not merely empty space; they are structured space. A floodplain forest, a marsh, a patchwork of shallow ponds and channels: these are rough landscapes that force water to do work on friction rather than on destruction. Vegetation acts like a drag net. Stems and trunks create turbulence that steals momentum. Even dead wood and leaf litter matter, because small obstructions multiply eddies. Where a concrete channel encourages a coherent, high-speed jet, a wetland encourages a messy, slow sheet of water. Messiness, in this context, is safety.

There is an emotional tension here. People often accept wetlands as wildlife habitat, scenery, or "nature." Asking them to accept wetlands as part of flood protection can sound like an attempt to replace serious infrastructure with sentimentality. The point is not that reeds replace levees everywhere. The point is that flood risk is often reduced more reliably by giving water somewhere to go than by insisting it stay within a narrow corridor whose failure, when it comes, is abrupt.

A concrete historical example sits in California's Sacramento Valley, where the problem was never a lack of engineering ambition. By the late nineteenth and early twentieth centuries, as farms and towns expanded, the Sacramento River system was increasingly hemmed in. Levees multiplied. Channels were straightened and dredged. The valley's rivers, fed by winter storms and Sierra snowmelt, repeatedly reminded residents that confinement has limits. Levees break. When they do, the water does not politely return to the channel; it pours into the lowest available space, often where people have built.

One of the most consequential decisions in the region was to *design* a place for flooding rather than treating flooding as an accident. The Yolo Bypass, west of Sacramento, is essentially a vast relief valve: during high flows, water spills into a broad floodplain corridor that can convey a large fraction of the basin's floodwaters. It is engineered, but it is engineered as *space*. In much of the year it supports agriculture and wetlands; in high water it becomes part of the river.

What makes that idea powerful is its honesty about scale. If a flood brings an enormous volume, trying to squeeze it through a single narrow channel is like trying to force an elephant through a doorway. You can reinforce the doorframe, but you are still arguing with geometry. The bypass solves the problem by changing the geometry. It provides a second, wide pathway where water can move with lower depth and velocity. When inundated, it also becomes a seasonal wetland landscape, producing food webs that fish and birds can use. That ecological benefit is not a decorative side effect; it is connected to the hydraulics. A productive floodplain depends on periodic flooding, and periodic flooding is exactly what attenuates peaks.

Levee setbacks follow the same logic on a smaller footprint. Instead of building a higher wall right on the bank, a setback moves the levee back from the river, widening the corridor. That widened corridor can hold more water during floods, lowering the stage (water surface elevation) in the main channel. It can also allow the river to deposit sediment on its floodplain rather than carrying it all downstream. A setback is, in effect, a partial admission that the river wants elbow room. The engineering is not eliminated; it is repositioned.

The difference between walls and sponges is also a difference between *local* and *system* thinking. A wall is local: it protects a particular parcel until it fails or until the flood exceeds its design. A sponge is systemic: it changes how the flood wave evolves through the network, sometimes reducing risk not only at the restored reach but downstream as well. But systemic effects are also why promises must be careful. It is easy to say "wetlands reduce floods" as if it were universally true. The truth is conditional.

Wetlands are diverse, and their flood influence depends on where they sit in the landscape and what state they are in when the storm arrives. A floodplain wetland connected to a river has obvious storage potential, because it can fill like a shallow bowl when the river overtops its banks. An upland wetland perched on saturated soils can behave differently. If the wetland is already full from earlier rain, it may contribute to runoff rather than reduce it. If it is drained by ditches, it may accelerate flow. Even in a floodplain, a wetland that has been cut off by levees is not available storage, no matter how lovely it looks on a map.

Antecedent moisture is the unglamorous key. A dry sponge absorbs; a saturated sponge drips. Rivers, soils, and wetlands have memory. A series of storms can prime a catchment so that the next storm produces far more runoff. In such cases, the question is not whether wetlands matter, but whether the available storage is large enough relative to the event. Small storage can make a meaningful difference in small floods and make almost no dent in an extreme flood.

That scale issue has practical consequences for "natural flood management" measures that have become popular in some regions: leaky barriers in small streams, re-wetting peatlands, adding woody debris, restoring hedgerows and small ponds. Many of these measures can slow flow locally, increase infiltration, and reduce erosion. They can also create valuable habitats and improve water quality. The temptation is to add them up in the mind like coins and assume they will buy protection from the largest floods. Sometimes they will not. A thousand small delays do not necessarily stack neatly into a big delay downstream, because flood waves merge, tributaries synchronize or desynchronize, and storage fills. The network is not a straight pipeline; it is a branching system with feedbacks.

None of this diminishes the value of wetlands and floodplains. It clarifies what they are good at and how they should be used. Floodplain reconnection is most convincing as a way to reshape the middle of the hydrograph: to shave the crest, slow the rise, and reduce erosive velocities. It is also a way to turn floods from purely destructive events into events that renew landscapes. Sediment deposited on floodplains can build soil. Nutrients spread across wetlands can stimulate productivity. Fish can move into shallow areas that are warm and food-rich. These are not sentimental benefits. They are the river performing its long-term work of maintaining a living corridor.

There is also a human benefit that is easy to overlook because it does not appear in a cross-section drawing: failure mode. Walls tend to fail abruptly. A levee holds until it does not, and then the energy stored in high water releases violently. Sponges fail gradually. When storage begins to fill, water spreads into designated areas, levels rise more slowly, and the "loss" is not a breach but an inundation that was anticipated. The difference between an emergency evacuation at 2 a.m. and a planned closure of a farm road in daylight is not a small difference in lived experience.

Designing with sponges rather than walls requires a different kind of maintenance mindset. Concrete requires inspection for cracks and undermining. Floodplains require monitoring of channels, vegetation, and sedimentation. Wetlands require attention to connectivity: are there culverts or embankments blocking flow? Has a channel incised so deeply that it no longer spills? Has a floodplain become so isolated that it no longer fills except during the most extreme events? The aim is not to keep the wetland unchanged, but to keep it *func-*

tioning as a place where water can slow down and spread out.

It also requires candor about tradeoffs. Making space for water can mean accepting seasonal flooding of some land uses. It can mean shifting protection from one area to another, which raises questions of fairness. A wall makes a simple promise to a narrow group: "you are protected." A sponge makes a broader promise that can sound less satisfying: "the system will be less extreme." That promise is harder to sell, even when it is more robust.

The most practical argument for wetlands and floodplains is not that they are morally better than concrete. It is that they exploit the same physics as concrete but do it with a broader palette. Concrete works by resisting water. Floodplains work by accommodating it. Resistance is expensive and brittle; accommodation is often cheaper and adaptable, provided the accommodation is planned rather than accidental.

A floodplain wetland, when connected to the river, is not just a pretty margin. It is a hydraulic device made of soil, plants, and space. It takes the sharpness out of the hydrograph by providing storage and drag exactly when the river is most energetic. That is what a sponge really is in river logic: not a soft thing, but a place where the flood can spend itself without taking homes and lives as payment.

9.3 Designing for Flux

A river network is a machine for turning uncertainty into motion. The uncertainty begins in the sky: where rain lands, whether it arrives as drizzle or cloudburst, whether snow piles up for months and then releases in a week of warm wind. Gravity translates that variability into flow, and the branching structure of a drainage basin gathers it into a few main stems. By the time the water reaches a town on a big river, the weather has been converted into a pattern of discharge that can be measured, forecast, insured, feared, and occasionally celebrated.

Engineers and planners have always worked with forecasts, but the more interesting change is cultural: a slow acceptance that "the design flood" is not a single number we can trust forever. A channel can be deepened, a levee raised, a seawall armored, and still the system can outrun its assumptions. Climate change alters storm intensity and snowmelt timing; urbanization hardens watersheds and speeds runoff; sediment trapped by dams or excavated from channels changes bed levels; sea level rise shifts the base level that rivers drain toward. Each of these forces changes the river's behavior without asking permission.

Designing for flux means treating variability as a permanent feature, not a temporary nuisance. It means using the language of hydrographs, probabilities, and network connections to decide where to resist water and where to accommodate it. It also means admitting that no single structure can "solve" a river. A river is not a pipe; it is a living corridor embedded in a society. The question becomes less "How do we stop flooding?" and more "How do we make floods less lethal, less surprising, and less unjust?"

The first tool is measurement: a habit of thinking in discharge rather than in vague impressions. Discharge sounds technical, but the idea is familiar. A faucet can deliver a fixed amount per minute; a river delivers a changing amount per second. The hydrograph is the river's pulse. When planners talk about a "100-year flood," they are not describing an event that happens once a century. They are describing an event with about a 1% chance of being exceeded in any given year, based on past data and statistical assumptions. In a stationary world, those assumptions might hold. In a changing world, the phrase can mislead, because it invites people to treat rare as impossible and to treat the past as a trustworthy guide to the future.

A second tool is topology: the geometry of connections in a network. In a river basin, topology is not an abstract math game. It is the practical question of which tributaries feed which towns, how quickly flood waves can synchronize, and where storage has leverage. Two communities can have the same rainfall but very different flood risk depending on where they sit in the branching structure. A town at the confluence of two flashy tributaries is exposed to coincidence: if peaks arrive together, the main river surges; if they arrive hours apart, the combined peak is lower. Small changes in land use upstream, or in the timing of snowmelt, can move those peaks closer together or farther apart. The network turns timing into height.

Topology also makes a mockery of purely local fixes. A levee that protects one reach can raise water levels downstream by pushing the flood wave onward with less friction and less spill. Channel constriction can increase flow velocity and scour, lowering the bed in one place but encouraging deposition in another where the river finally slows. Training works that pin a channel for navigation can reduce lateral movement but concentrate energy and raise stages during large floods. These are not moral failures; they are network effects. The river is a connected system, and the water does not forget what happened upstream.

Historical riverscapes offer countless examples of well-intended control that gradually tightened the system until the system broke. The Mississippi River is the most famous American case, not because it is unique, but because its scale makes the logic hard to ignore. After the catastrophic flood of 1927, the United States embarked on an ambitious program of levees, floodways, and channel improvements. The "levees-only" approach was already under criticism, and the later system did include designated spillways, but the dominant impulse remained: keep the river in its corridor and move floodwater downstream efficiently. Over decades, as levees grew higher and more continuous, the river's connection to its floodplain diminished. In places, bed levels rose because sediment that once spread across floodplains was forced to stay in the channel. In others, the river deepened and scoured, complicating navigation and infrastructure. The details vary by reach, but the overarching lesson is consistent: constrain a big river over long distances and you change the stage-discharge relationship. The same discharge can produce a higher stage. Protection can erode not through a single dramatic failure but through a slow upward creep of flood heights against static defenses.

That creep is why designing for flux cannot be reduced to "nature good, concrete bad." Sometimes a wall is the only responsible choice: around a hospital, a drinking-water plant, a dense historic neighborhood that cannot relocate. The question is where walls should end, where they should be set back, and what else must accompany them so they do not merely transfer danger.

Levee setbacks are a good example of a hybrid approach that makes network thinking concrete. A setback does not abandon flood protection; it rearranges it. By moving the levee away from the channel, the river corridor widens, allowing more conveyance and storage during high flows. The immediate hydraulic effect can be a reduction in water surface elevation for a given discharge near the project reach. The broader effects can extend beyond the reach, because changing flow depth and friction alters the propagation of the flood wave. But the benefits are not linear. Doubling the setback distance does not necessarily double the reduction in stage. In some settings, a small setback does little; in others, it crosses a threshold and suddenly creates meaningful storage and roughness. River valleys have geometry, and geometry produces tipping points.

There is also a less comfortable truth: setbacks can redistribute risk in ways that look like progress on a regional map but feel like loss at a kitchen table. If a new alignment leaves some homes outside the protected area, those homes become more exposed to frequent, smaller floods, even if the region is better protected against catastrophic floods. The mathematics of expected annual damages can favor a setback, while the lived experience of one unprotected neighborhood becomes worse. This is not a reason to reject setbacks; it is a reason to pair them with social policy: buyouts, relocations, and land-use changes that reduce exposure rather than pretending exposure does not matter.

A concrete illustration of this pairing can be seen in many modern flood projects that blend structural changes with land acquisition. A setback that widens the corridor may require purchasing land or easements. That purchase is not an administrative detail; it is the core mechanism by which risk is reduced. When society buys space for the river, it is buying a change in the hydrograph and buying the option for the river to spread out without turning into a lawsuit.

The institutional language around these projects has also shifted. In the U.S. Army Corps of Engineers, the phrase "Engineering With Nature" has become a banner for approaches that intentionally combine flood risk management with ecological function. The phrase can be used vaguely, but its best versions are quite strict: do not add "habitat features" as decoration; alter the system so that ecological structure emerges from hydraulic and sediment processes. A setback that creates seasonal inundation habitat while reducing stages is not an environmental add-on. It is a single intervention doing two jobs because it is aligned with how the river wants to behave under flood conditions.

Forecasting becomes the nervous system of this approach. The point of better discharge data is not merely to issue flood warnings; it is to decide how much room to give the river and when. In basins with reservoirs and bypasses, managers can sometimes shape the flood wave deliberately: releasing water earlier to create storage, routing flows into designated floodways, timing releases to avoid peak synchronization at downstream confluences. Even where such operational control is limited, forecasts allow communities to shift from reactive to anticipatory behavior: closing floodgates, moving vehicles, protecting critical equipment, evacuating vulnerable residents.

But forecasting alone cannot solve a problem of deep uncertainty, where future conditions may depart substantially from the past. Deep uncertainty is not just "we don't know the exact number." It is "we are not sure which model of the future is right." Will extreme rainfall increase by 10% or 40%? Will snowpack decline enough to change the seasonality of floods? Will a riverbed aggrade (build up) due to sediment or degrade (cut down) due to incision? When those questions are open, a single, fixed master plan becomes a gamble.

One response has emerged from the planning world: Dynamic Adaptive Policy Pathways (DAPP). Behind the bureaucratic name is an everyday logic. Decide what outcomes matter (for example, a maximum acceptable probability of flooding in a neighborhood). Identify *tipping points*: conditions under which the current strategy fails to meet those outcomes (for example, a certain sea level rise, a certain increase in peak discharge, a certain rise in river stage for a given flow). Implement actions that work well now and keep options open. Monitor *signposts*: measurements that tell you whether you are approaching a tipping point (changes in extreme rainfall statistics, observed trends in flood stages, subsidence rates, sedimentation). Preplan what you will do when the signposts suggest the pathway is becoming unsafe.

This way of thinking treats a flood strategy as something like a medical plan for a chronic condition rather than a one-time surgery. The river changes, so the plan has to change. Crucially, the plan is not improvised during a crisis; it is mapped in advance so that switching strategies is politically and logistically possible.

Flux design also forces a more honest conversation about what counts as "protection." A high levee can reduce the frequency of flooding and thereby encourage development behind it. Development increases the value at risk. When an extreme flood finally exceeds the levee, the damages can be far larger than they would have been if the land had remained less developed. This is not merely theoretical; it has been observed repeatedly in protected floodplains worldwide. The protection changes behavior, behavior changes exposure, exposure changes consequences. The flood risk becomes a social feedback loop as much as a hydrologic one.

Breaking that loop requires land-use decisions that feel, at first, unrelated to rivers: zoning, building codes, insurance, and the willingness to say "not here." A river corridor can be engineered and restored and still remain dangerous if society insists on placing the most vulnerable housing and essential services in the lowest ground. Conversely, a society can tolerate a great deal of hydrologic variability if it is careful about what it places in the flood's path.

Topology returns here, not as a property of water but as a property of people. Roads, power lines, evacuation routes, hospitals, and supply chains form networks that can fail when certain nodes are flooded. A small inundation at a critical junction can have outsized effects. Designing for flux means identifying those nodes and protecting them disproportionately, while allowing less critical areas to flood in controlled ways. It is a form of triage that can sound cold until one remembers that pretending everything can be protected equally is often what produces the worst outcomes.

A useful way to visualize the goal is not "zero floods" but "bounded floods." Bounded floods happen in places chosen in advance: parks designed to fill, agricultural fields under easement, restored floodplains behind setback levees, bypass corridors with controlled inlets. Bounded floods are legible to residents: water goes where it is supposed to go, at roughly the times and depths that were anticipated. The hydrograph still rises, but the peak is shaved and the consequences are distributed more gently.

This is also where the earlier emphasis on roughness and space becomes political. A wetland that attenuates peaks is not just habitat; it is a component of a risk-sharing agreement. Someone, somewhere, is agreeing that their land will hold water so that someone else's living room does not. Making that agreement fair requires compensation, transparency, and often a change in how communities value certain landscapes. The floodplain stops being "empty land waiting for development" and becomes functioning infrastructure, even if it remains green.

There is no single blueprint for what a resilient river society looks like, because basins differ in climate, topography, wealth, and history. Still, a pattern emerges whenever the work is done well. The river is given space where space is cheapest and most effective. Hard defenses are concentrated where the stakes are truly irreducible. Forecasting is treated as a continuous public service rather than an occasional alarm. Adaptive plans are built around signposts and tipping points rather than false certainty. Exposure is reduced not only by walls but by choices about where people live and what they build.

A river's deepest lesson is not that water is powerful. Everyone knows that. The lesson is that power does not disappear when constrained; it reappears elsewhere in the network, sometimes after a delay long enough for complacency to grow. Designing for flux is the habit of staying ahead of that reappearance: letting water move and spread where it can do little harm, while protecting what must be protected with the sober awareness that the river is not a static object but a changing relationship between landscape, climate, and society.

www.ingramcontent.com/pod-product-compliance
Lightning Source LLC
LaVergne TN
LVHW022206200926
848481LV00006B/1869

* 9 7 9 8 8 9 6 6 5 2 7 0 0 *